Citizen Science Fiction

Citizen Science Fiction

Jerome Winter

LEXINGTON BOOKS
Lanham • Boulder • New York • London

Published by Lexington Books
An imprint of The Rowman & Littlefield Publishing Group, Inc.
4501 Forbes Boulevard, Suite 200, Lanham, Maryland 20706
www.rowman.com

6 Tinworth Street, London SE11 5AL, United Kingdom

British Library Cataloguing in Publication Information Available

Library of Congress Cataloging-in-Publication Data

Names: Winter, Jerome, author.
Title: Citizen science fiction / Jerome Winter.
Description: Lanham : Lexington Books, [2021] | Includes bibliographical references and index.
Identifiers: LCCN 2020057374 (print) | LCCN 2020057375 (ebook) | ISBN 9781793621474 (cloth) | ISBN 9781793621481 (ebook)
Subjects: LCSH: Science--Study and teaching. | Research--Citizen participation. | Science--Social aspects. | Science fiction in science education.
Classification: LCC Q181 .W754 2021 (print) | LCC Q181 (ebook) | DDC 507.1--dc23
LC record available at https://lccn.loc.gov/2020057374
LC ebook record available at https://lccn.loc.gov/2020057375

Contents

Introduction

"Don't do it, Citizen! Littering the streets is an offense!" Thus shouts the staunch future-cop Judge Dredd in the "Full Earth Crimes" (1978) comic-strip story line, scripted by John Wagner and penciled and inked by Brian Bolland and Mike McMahon, first serialized in the influential magazine *2000 AD*. Retaining an essential superheroic core beneath his grim antiheroic exterior, Judge Dredd yells this absurdly ecofascist warning as he valiantly swoops down in his flying motorcycle to rescue a suicidal jumper who has flung himself in despair from a towering overcrowded skyscraper. The stone-faced and forever behelmeted Judge Dredd then subsequently charges the mentally ill man with being a public nuisance and harshly books him for ninety days of penal servitude, thereby maintaining his no-nonsense fearsome reputation as the most merciless law enforcer in the postapocalyptic metropolis known as Mega-City One. Rhetorically, this pulp-infused hailing of the law-abiding "citizen" as an honorific epithet of dystopian propaganda suggests that the cultural idea of citizenship cannot be neatly divorced from biopolitical discourses that privilege the policing and surveillance of state sovereignty as Hobbesian hedge on the war of all against all.

Indeed, some academic critics contend that citizenship as an abstract category amounts to an irredeemably invidious distinction concocted by the modern capitalism-driven nation-state that discriminates against and arbitrarily excludes individuals from basic access to necessities of survival. Dimitry Kochenov, for instance, eloquently castigates the celebration of good citizenship as an inherent democratic right and responsibility: "abstract and essentially impersonal, citizenship is totalitarian in nature: it does not emerge in 'dialogue' and is much less flexible than many would like to think" (38). Against this problematic construction of citizenship as a toothless variety of civility politics, many others argue that the overriding mission of educational institutions should be, as Martha Nussbaum describes the imperative, to produce and cultivate thoughtful citizens more broadly conceived as "the ideal of a citizen whose *primary* loyalty is to human beings the world over, and whose national, local, and varied group loyalties are considered distinctly secondary" (9). Given this contested linkage between sovereignty, citizenship, and education, it is therefore perhaps unsurprising to find a compellingly nuanced formal definition of "citizen science" issued by the United States Congress in The Crowdsourcing and Citizen Science Act of 2016:

> The term "citizen science" means a form of open collaboration in which individuals or organizations participate in the scientific process in various ways, including (A) enabling the formulation of research questions; (B) creating and refining project design; (C) conducting scientific experiments; (D) collecting and analyzing data; (E) interpreting the results of data; (F) developing technologies and applications; (G) making discoveries; and (H) solving problems.

By no means a totalitarian bill, this innocuous piece of legislation formally grants the federal government and states the authority to sponsor, fund, and coordinate citizen-science projects, stipulates protections of privacy for participants, declares the official intention for citizen-science data to be made public and open-sourced, sets ups guidelines for partnerships between market and nonprofit organizations in the administration of citizen science, and creates an accessible toolkit and website to assist in the development of citizen-science projects. The congressional bill might be seen, then, as officially sanctioning what had already become readily apparent to the previously converted legions of citizen-science acolytes, aficionados, and professionals who had been busily amassing their numbers for decades, if not centuries. The bill, after all, is only the next logical extension of the recent rapid expansion of citizen commitment to the operations of science, a technologically assisted transformative shift in the relationship between scientific communities and the public sphere, laypeople, and experts. The bill endows citizen science with an aura of public trust, confidence, and credibility it has already widely begun to wield in the realm of local and global cultural politics writ large, further legitimating the phenomenon as a significant restructuring of the social networks, institutions, methodologies, and apparatuses that constitute how ordinary people understand and get involved in science today.

In the educational context, however, the dystopian, exclusionary connotations of citizenship in citizen-science rhetoric and discourses have not gone unnoticed. A 2018 National Academies of Science Report, *Learning through Citizen Science*, following in the footsteps of the congressional act, frankly expresses the concern that "the term 'citizen,' particularly in the United States, connects to a contentious immigration debate about who is eligible to participate in civic life, including science and education" (14). Indeed, as a devastatingly frank critic might put it, citizen science can all too easily be characterized as overworked and underfunded scientific experts, experimenters, and authorities deploying manipulative guile to dragoon unpaid menial labor from entitled, naïve dilletantes who have the privileged luxury of flaunting their civic status and donating their precious leisure time to their glorified hobbies. Without ever dismissing this perhaps unduly fierce critique out of hand, however, *Citizen Science Fiction* argues in contrast that citizen science might be viewed as a remarkable evolution of crowd-sourced projects, an intensification of a globally affiliated network of cosmopolitan citizens, and a sociological sea change in the scientific establishment. This social and political transformation promises to dismantle some salient cultural barriers that blockade laypeople from scientific expertise and unproductively entrench pro-science from antiscience ideologues in cultural exchanges. Even so, it must continually be remembered that citizen science, even when uncomfortably draped in outwardly benevolent or progressive rhetoric, is not merely equivalent to other noble acts of civic engagement, such as volunteering in a soup kitchen, donating blood, marching against a social injustice, or canvassing for a local election.

Citizen science has become affiliated with many acronyms and keywords in the proliferating academic literature on the subject, such as participatory action research, community-based management, and public participation in scientific research. However, citizen science is what the phenomenon is usually referred to in colloquial conversation, perhaps because a companionate marriage between the everyday understanding of citizenship and science is integral to the phenomenon. This book *Citizen Science Fiction* continues the precedent set by James Wynn's *Citizen Science in the Digital Age* (2017), which invokes the Aristotelian ethical-rhetorical terminology of "good judgment" (*phronesis*), "virtuousness" (*arête*), and "goodwill" (*eunoia*) to segregate the consumer hobby of regurgitating beer-mat fun facts and ventriloquizing mass-media science popularization, or what Wynn identifies as "semi-expertise" or "association to expertise," from more substantial "interactional" and "contributory" citizen-science expertise in which one can witness "interaction with the culture and contribution to the production of science that will guide the theoretical distinction . . . between technical and nontechnical ethos" (71). As discussed in chapters that follow, the plentiful examples of citizen science studied in the book and validated as credible and qualified even by authoritative expertise achieve this objective and transparent methodological or technical ethos by doing the hard work of real science: generating research questions, testing hypotheses, assessing risks, demystifying expertise, and making policy. Readers will recall that The Crowdsourcing and Citizen Science Act of 2016 defined citizen science as aspiring to transmitting such interactional and contributory technical ethos to its eager and accomplished participants. Nevertheless, what this book wishes to add to Wynn's astute inquiry into the Aristotelian rhetoric of citizen science is that the putatively objective and neutral purity of the technical ethos additionally relies on a coproduction of scientific credibility and an intricate matrix of historical discourses, cultural politics, and social norms ripe for integrating into rhetoric-and-composition pedagogy.

In dark times marked by posttruth disinformation campaigns and the viral dissemination of false media endorsed by autocratic rulers, pro-science apologists, committed popularizers, and earnest activists frequently assume that science-deniers, the misinformed, and fact-ignorers spread their falsehoods out of apathy, a lack of education, or an ideological imp of the perverse. Yet recent research shows that conspiratorial true believers often construct their byzantine arguments out of a spirit of vigorous misconstrual, self-deception, and willful, self-aware defiance. Hence high-handed correction and censorious remonstration can prove soundly unpersuasive and counterproductive to these ideologically entrenched contingents deeply influenced by antiscience agitprop. Such flawed science communication unconsciously adopts the deficit or banking model of learning that views science as transparent, monolithic, and neutral, and the general public as intractably prone to passive paths of least resistance, ignorant error, and irrational frenzy. Moreover, the frustrated cries, dismissals, and heated aspersions lobbed by militant science apologists against the irresponsibility, hysteria, and flagrant ignorance of blinkered citizens are also too often directed, problematically and unhelpfully, by overwhelmed and exhausted traditional educators against their struggling students.

Likewise, rather than the dystopian invocation of citizens as either fractious or docile subjects waiting to be rectified with the righteous doling out of enlightened reason, the thick mesh of responsibility, obligation, pleasure, curiosity, and community implied in the participatory ethos of citizenship as conceived by citizen science should be understood along the lines demarcated by Cory Doctorow's science-fiction (sf) novel *Walkaway* (2017). Doctorow writes, "You never hear politicians talking about 'citizens,' it's all 'taxpayers,' as though the salient fact of your relationship to the state is how much you pay" (67). Against narrowly conceiving citizenship as a state-sponsored corporate loyalty program, Doctorow imagines a critically utopian citizenship as motivated by the inherent resourcefulness, keen intelligence, and active generosity of a vibrant civil society energized by science-driven collaborative projects. If not exactly anarcho-syndicalist, citizen science might still therefore be reconstructed as technologically advanced, do-it-yourself enclaves of proactive people collaborating in impromptu direct democracy or decentralized found families. This reinvigorated civil society can help mitigate the widespread ecological and economic disaster bred by the reactionary special-snowflake delusion of extreme individualism rampant among state-sanctioned titans of industry and neoliberal lawmakers.

Naturally, science is not a uniform entity from which one can easily walk away or with which one can categorically identify, and neither should citizen science be conceived as a monolithic concept before which we must genuflect. In *Citizen Science* (1995), Alan Irwin inaugurated the academic study of citizen science by closely reading the resistant and subversive civic engagement in the agricultural use of pesticides, the consumption of mad-cow meat, and toxic runoff of chemical sites. Irwin argues that in these disparate cases citizen science has endeavored to reclaim, reconfigure, and radicalize scientific institutions to address and respond to the needs and values of the vulnerable public sector and directly affected populations more effectively. Irwin imagines the popular reception of the scientific establishment might be altogether reimagined then not as an "unstoppable juggernaut" but as a "best friend," and the shared governance of engaged citizens as a counterhegemonic public "who is prepared to offer unpalatable advice" (2). A more recent example of Irwin's notion of citizen science might be found in the widely publicized collaborations between the concerned citizens of Flint, Michigan, and the engineering professor Marc Edwards at Virginia Tech, who collectively measured the toxic levels of lead in tap water after LeeAnne Walters began noticing skin irritations and hair thinning at home.

An alternative but compatible model of citizen science can be found in Rick Bonney and Janis Dickinson's *Citizen Science* (2012) that underscores that citizen science instills in a sophisticated and gracious public the fun-filled pleasures of curiosity-driven enthusiasm, the joys of empirical discovery, and the personally motivating power of self-fulfillment. Faced with widespread legitimation struggles and disinformation, citizen science becomes a useful tool to help recruit laypeople, amateurs, and the nonprofessional public to regain confidence in the construction of credible scientific knowledge. Hence Bonney and Dickinson hope that citizen science will be instrumental in the expansion of scientific literacy across a wide spectrum of society, and "habits of scientific inquiry will develop to promote scientific decision making from a position of knowledge and skill rather than of powerlessness" (10). A major example of this empowering vision of citizen science may be drawn from the Kyoto Protocol that urged in 2005 that governments around the world immediately respond to climate change based in part on datasets collected from passionate citizen-science bird-watchers who collectively observed birds unseasonably laying their eggs earlier in the year than normal.

The extraordinary profusion and remarkable multiplicity of citizen-science projects that have proliferated in recent years can be witnessed by visiting the SciStarter online portal. Created by Darlene Cavalier and affiliated with the Arizona State University's School for the Future of Innovation in Society with support from the National Science Foundation, SciStarter offers a free, customizable, and searchable dashboard that hosts profiles, portfolios, reviews, blogs, podcasts, message boards, and other gamified elements to connect citizen-science project managers and scientists with citizen-science volunteers and educators. The SciStarter portal features thousands of citizen-science projects ranging from bird-watching, beach walking, and stargazing to the tracing of neurons, the testing of water, the folding of proteins, and the marking of language variability. Cavalier, though, repeatedly emphasizes that citizen science should be distinguished from other educational outreach campaigns geared toward teaching Science, Technology, Engineering, and Mathematics (STEM), or science-literacy advocacy conceived more generally.

Cavalier contends that citizen science specifically amplifies the public understanding of science by critically rethinking who can actively participate in constructing the technical ethos of scientific knowledge; therefore, truly transformative citizen science, according to Cavalier, views the citizen scientist both as a robust instrumental resource and empowered agent, "elicit[ing] knowledge from citizen scientists to advance scientific research or to solve a real problem" (Cavalier et al. 35).

Consequently, the rhetoric of citizenship inherent in citizen science should not be extricated from its ordinary cultural context and treated as an arbitrary, free-floating signifier with unfortunately draconian associations. Instead, radicalizing citizenship as an emancipatory pedagogical and civil priority should be reclaimed to counterbalance the authoritarian impulses of uncritically valorizing elite technocratic expertise. Sophisticated Marxist critics have long acknowledged that the technological capital accumulation of the political economy couches itself in civil society as much as the umbrella ideological superstructure of legal codes, cultural institutions, and political organizations. Likewise, artistic, scientific, and academic modes of production can be traced back in the last analysis to an economic base of inequality, commodity fetishism, social injustice, and alienating class struggle. Terry Eagleton, for instance, notes that in his advocacy for revolutionary change, Karl Marx himself pointedly did not disparage the gradual and progressive reform achievements of bourgeois civil society, championing "socialism as the inheritor of its great legacies of liberty, civil rights and material prosperity" (239). Correspondingly, the widespread lack of diversity and equity in scientific institutions and organizations derives from minority and disadvantaged students understandably viewing scientific curriculum as remote, esoteric, and irrelevant to their everyday experiences, struggles, and lives. Hence *Learning through Citizen Science* argues that citizen science must be commandeered to reverse the systemic racism, sexism, and class bigotry of our civil society and to "question embedded and pernicious assumptions about who is capable of participation and what that participation may yield" (6). This call extends years of consciousness-raising pedagogy research pioneered, for instance, by bell hooks, in *Teaching to Transgress* (1994), who powerfully argues that harnessing the cachet of academic discourse for the catharsis of a participatory and engaged civics-oriented classroom can be meaningfully liberating for students provided that student voices, experiences, and struggles are continually foregrounded such "that the knowledge received in these settings will enrich and enhance them" (19).

Injecting student-driven grappling with issues of citizenship into the study of science might be rejected as anathema by traditional scientists themselves who publicly avow the neutral, transhistorical truth value of pure scientific investigations. The problem is analogous to the dilemma of another sf novel, China Miéville's *Embassytown* (2011), in which a kleptocratic galactic empire seeks to tighten its dominion over its cosmic frontier, especially the trading outpost world, Arieka, whose indigenous alien Hosts, baffling "insect-horse-coral-fan-things" (121), possess such metaphysical and physiological alterity that they speak, from two orifices simultaneously, what they call simply "Language," which forbids metaphorical untruth except by extreme contortions. The accidental introduction of falsehoods into the Hosts' Language, by a botched clone experiment conducted by the empire, ushers in waves of social and political unrest, chaotic mass slaughter, and the eventual insurgent transformation of the language spoken by the Arieka into both instances of valid referentiality and "some indispensable brew of contradiction, insinuation, and untethered meaning" (171). Ironically, Miéville's prototypical sf novel exemplifies Darko Suvin's epochal definition of sf literature as "cognitive estrangement," or the dehabituation and sensitization of readers to socially progressive and technologically utopian aspects of scientific and Enlightenment worldviews. This cognitive effect is ironic since in his academic criticism Miéville himself deconstructs Suvin's prescriptive definition of cognition as overly reliant on the "charismatic authority" (Miéville, "Afterword" 238) of scientific and technological expertise. Miéville's unfathomable, literal-minded aliens free fall into the overdetermined and polysemous state of ambiguous language. Likewise, what counts as authoritative cognitive rationality in science, let alone the subsidiary notion of cognitively estranging sf literature and media, depends on a plural, mutable, and complex system of unstable signification, the slippages and ruptures of untethered cultural semiotics and rhetoric. As John Rieder argues about the Suvinian paradigm of sf genre theory but with a compelling logic that we might also extend to understanding what scientific writing means for the public sphere more generally, scientific expertise constructs the pure and objective verisimilitude of science by "acts of definition, categorization, inclusion and exclusion (all of which are important), but also by their uses of the protocols and the rhetorical strategies that distinguish the genre from other forms of writing and reading" (38).

The overarching contention of *Citizen Science Fiction* is that student research and writing can be guided, intervened in, and, most importantly, encouraged not by the wholesale adoption of a mutually reinforcing combination of citizen science and sf literature and media into a specific, prescribed, or canonical curriculum. Rather, a necessarily eclectic, motivation-spurring combination of sf and citizen science, as mutably and flexibly suggested by specific case studies in this book, can be selected, adapted, modified, compacted, and extended at will for the improvisational pedagogical needs and ad-hoc priorities of writing instructors on the job. In line with recent trends in rhetoric and composition theory, this book provides discussions of the content of citizen science projects and sf works in tandem then only as concrete illustrations, detailed recommendations, and specific curricular possibilities. The primary preoccupation of *Citizen Science Fiction* therefore remains focused on contributing to the theorizing of reading sf-infused citizen science as pedagogical tools for stimulating class discussion, sparking academic and cultural engagement, and helping students learn how to write about what they are often by and large already deeply interested in—namely, challenging issues of science and technology that directly affect their everyday lives and futures.

The perennial debate over the place of reading and writing in education has historically served a legion of divergent masters, yet recently renewed scholarly interest has gravitated toward an intensified focus on the proper place of reading in composition pedagogy. Generally, Ellen Carillo delineates the overarching trajectory of composition theory as tending toward the consensus that "the range of activities that falls under what might be called 'reading' demands a more complex practice than a onesizefitsall mechanical process of decoding" (7). Dominant in primary and secondary educational practice as well as teaching English as a second language, an emphasis on decoding, or the neuro-physical translation of phoneme and graphemes to meaningful units of comprehension, demands what Kathleen McCormick calls an objective or empirical "cognitive model" (14) of pedagogical approaches to teaching reading.

As opposed to the remedial acquisition of the cognitive skill of decoding, basic-comprehension reading strategies, as codified by Cathy Block and Gerald Duffy, and often incorporated into more advanced literacy standards, require explicit scaffolding, direction, and modeling of specific reading strategies, including the following goal-oriented reading plans: predicting, monitoring, questioning, imagining, inferring, summarizing, evaluating, and synthesizing (22). As frequently emphasized in First Year Composition (FYC) courses, these reading-comprehension strategies and plans, though, depend on the more fundamental workings of what we might call, following Robert Scholes who himself adapts the phrase from a critical reading of Jacques Derrida, "reading protocols" (Scholes 88). In terms of the practical repertoire of the writing classroom, the reader-response hermeneutics of protocol theory can be understood as expressed in the kind of essayistic and dialogic inquiry pioneered by David Bartholomae and Anthony Petrosky as "ways of reading," which entails encouraging student responses to reading in writing assignments that are "tentative, open to the possibility of elaboration, modification, or revision through further dialogue and ongoing reflection on the text and/or other's reaction to the text" (Qualley 66).

An emphasis on teaching reading protocols may eschew both lapsing into excessive concentration on rudimentary drills that reinforce cognitive skills, honing unreflexive exercises that merely practice basic literacy strategies, or a numbingly overawed deference to simply appreciating the almighty text or author as occurs in some literature curriculum. Instead, reading protocols view reading processes as dynamic, inquiry-driven springboards that simultaneously bring to bear and shape the variable and pluralistic examples of student experience, knowledge, and engagement. Nancy Morrow, for instance, articulates a fascinating catalogue of reading protocols to refine and complicate the strategies and skills previously mentioned as well as to supplement the most commonly applied protocols in college composition, such as reading for genre convention and reading for persuasion. Morrow's more innovative, exploratory reading protocols include reading to develop an intellectual repertoire, reading for parsing ambiguity, reading for highlighting the unexpected, and reading for linguistic play (467–69).

However, an overlooked aspect of the implicit reader-response theory of this educational praxis, as Jane Tompkins observes, is the half-buried, underlying ethical and social imperatives that inform this reading pedagogy. Whether these imaginative reading protocols assist writing instruction that is traditional and content-based, formalist and structure-based, or poststructural and process-based, or more likely, all of the above, all theoretical approaches depend on the assumption that cultivating the classical idea of "virtuous action" (xv) lies at the heart of the reading and writing enterprise. Far from new to teachers of composition and rhetoric, approaching reading as furnishing models for a broadening of personal horizons and enriching of moral experience, of course, has its ancient roots in the rhetorical practice of imitation (*imitatio*), which, in contrast to automatic regurgitation or perfunctory emulation, has historically underscored "invention and revision, expression and discovery, cognition and collaboration" (Farmer and Arrington 12).

A nominal consensus also seems to have been reached that teaching the "how" of reading skills, strategies, and protocols can be variously applied to a wide variety of specific "what" or assigned reading, regardless whether the text selection under classroom scrutiny may be considered difficult canonical literature, a fun, disposable airport-terminal potboiler, or a specialized piece of technical writing. In the careful elaboration of a position that has now become perhaps mostly noncontroversial, Gerard Graff makes the case, for instance, that interrogating "weak" texts—that is, lightweight consumer artifacts of popular media and entertainment—can expose and socialize students to the "folkways and norms" of academic and professional discourse as much as an expert exegesis of a "strong" text, that is, a highbrow literary work of esteemed imaginative fiction or creative nonfiction. In scholarship on reading, this binary between teaching strong and weak texts in freshman composition has become tightly coupled to the debate between Erika Lindemann and Gary Tate in the March 1993 issue of the journal *College English*.

There is certainly no need to relitigate the frequently cited and decades-old Lindemann/Tate exchange here; suffice it to note that their diametrically distinct perspectives, and the corresponding dichotomy between reading-driven and writing-oriented pedagogy, or what, during the full waging of the culture wars it may be noted, was once called the rival camps of literature and composition, are by no means irreconcilable. Note, for instance, that in making his passionate case for teaching of a strong text of imaginative literature in the freshman composition course, Gary Tate contends that such work is stimulating for undergraduates precisely because failing to teach the expected literature would deprive students of lively conversations about what is commonly perceived as the eternal verities of life, and therefore sadly refuse to satiate the vital student hunger for "conversations [that] take place *outside the academy*, as they struggle to figure out how to live their lives" (320, emphasis in original). Likewise, in making her case against robbing students of a primary focus on the writing process itself through a lecture-oriented detour into a recondite graduate-student or literature-major analysis, such as tracing "the ingrown toenail motif in *Beowulf*" (314), Lindemann contends that the dichotomy between "humanists" and "scientists," or between the "academy" and "'*real life*'" (315–16, emphasis in original), has been greatly overstated. Regardless of the instructor's ultimate pedagogical choice of teaching literature or not, both Lindemann and Tate tacitly agree here that teaching reading protocols and rhetoric in a composition classroom can deploy both weak and strong texts, both classic literature and professional prose, to furnish students with rich material to reflectively meditate on their own personal life experience as well as gradually accumulating prior knowledge, which will necessarily involve critical engagement with a broad array of civic discourses, social norms, and public values. Whether students find themselves more engaged with readings that primarily approach citizen science fiction through a focus on citizenship, citizen science, science, or science fiction, or all the above, carefully integrating reading on this broad, inclusive topic into the writing classroom can serve as rocket boosters for their writing.

Anticipating the standards-driven sway of informational texts and expository writing in our current secondary-education pedagogical environment, the legacy of Kenneth Burke stands as a useful antecedent for contemporary composition studies in theorizing reading protocols when he opposes an emphasis on collaborative rhetorical dialogue, and its furnishing of equipment for living, to the naïve positivism and empty empiricism of popular scientific discourse. Wayne C. Booth was perhaps one of the first of Burke's keen readers to notice that such innovative openness to multiple perspectives when discussing informational and scientific texts helps articulate the expansive power of critical reader responses, engagement, and deconstructions such that "even scientific pursuits have a place under his own kind of pluralistic umbrella" (*Rhetoric of Rhetoric* 106); indeed in the seminal text *The Understanding of Scientific Prose* (1993), edited by the rhetoric-and-composition scholar Jack Selzer, which collects together a rich profusion of responses to a single scientific paper by Stephen Jay Gould, Burke's oxymoronic notion of the rhetoric of the "human sciences" is invoked as a crucial "forebear" (4) that anticipates what would later be consolidated by diverse scholars as major works in the philosophy of science, Science and Technology Studies, (STS) the Social Construction of Technology (SCoT), and technocultural studies. Burke also anticipates the flourishing academic field of the rhetoric of science, and the rich and variegated work that this book will regularly draw on, including scholarship by rhetoricians Leah Ceccarelli, Michael Overington, Jeanne Fahnestock, Alan G. Gross, Karen Schroeder Sorensen, and Thomas Lessl.

In "Four Master Tropes" (1941), Burke probes "the limits of science, qua science" (423). By contrast to "poetic realism" or the essential hypostatizing of ideational substance and being, Burke suggests science posits the study of a dynamic flux of material-physical conditions, processes, and correlations, labeling the rhetoric of this research "scientific realism" (422). However, even scientific realism, Burke avers, depends on an unacknowledged system of rhetorical symbolism and tropes, such as—to use some rhetoric discussed in the upcoming chapters of this book—metonymy (as in the image of the double helix of DNA), metaphor (as in the idea of a thinking machine), synecdoche (as in the greenhouse gas effect), and irony (as in the dialectical case for and against manned space exploration). Ultimately, Burke indicts what he elsewhere calls "scientism" as the doomed impulse to extricate the putatively pure and neutral abstractions of science from the unruly chaos of cultural discourse, rhetorical gestures, symbolic layering, and poetic modes of cognition. Informed by diverse reading protocols, civic education and a citizen readership of scientific discourse, becomes deeply invested in fostering dialogue, extending conversations, and encouraging exchange within a plural, mutable, and fluid culture. This dialectical dismantling of strict scientism may provoke rhetorical engagement that challenges the overweening prerogative of scientific expertise, subverting the putative neutrality, objectivity, and independence of reified scientific inquiry. Reading protocols that excavate the hidden rhetorical richness of scientific discourse show how science communication revolves around its own rhetorical gambits, semiotic perspectives, symbolic logic, and theoretical assumptions. In a writing classroom, critical reading protocols can productively probe the implications of ambiguities, elisions, and misprisions that scientism, buffered by its immense cultural cachet, deigns to pass over in silence.

While by no means serving prescriptively as the only means to engage students, due to some formative contingencies of its historical makeup, the sf genre in particular has accidentally become as much a "how" of reading protocols as a "what" of text selection for college composition in its evolving status as a vibrantly accessible vehicle for negotiating the host of problematic issues broached by science and technology at large. Whereas recourse to plumbing some of the very best of what the sf genre has thought and said does not resolve the unending debate over assigning soft or hard texts in writing classrooms, or focusing primarily on teaching reading or writing for that matter, it nevertheless offers a possibility for productively merging questions of both how to write and what to read, and whether weak or strong texts should be preferred, since well-crafted sf works can appear superficially to be entertainingly weak but also often withstand strong interpretative scrutiny, rigorous analysis, and writerly unpacking. In the writing classroom, keying students onto the ongoing debate over the rhetoric of science that genre sf continually elicits in its devoted core of avid readers can at its best produce in otherwise distracted, overwhelmed, or exhausted students a transgressive cognitive estrangement of their immediate science-shaped historical circumstances.

In addition to analyzing the how and what of sf texts in general for their rhetorical inquiry and social perspectives on science, the precise response-based reading protocols this book would like to contribute to the ongoing scholarly discussion sketched above are reading for calibrating psychological mind-set and mindful argument, reading for probing inquiry and productive uncertainty in the age of the Anthropocene, reading for voice with a view to our digitally dominated future, and reading for threshold concepts in a scientifically driven society. Although specifically viewed through the hybrid combination of weak and strong textual discourse found to be so replete in sf literature and media, the assigning of such texts does not presume that sf serves merely as a handmaiden to science education or, alternately, as exhaustive studies of particular thematic terrain or prescriptive forms of literary or cultural expression. In contrast, my analytic spotlight on sf literature and media not only plays to my own particular expertise—always an important factor in teaching composition—but also hopes to reveal the extent to which such material can serve as illustrative case studies for deeply interrogating civic engagement with science and its incorporation into a rigorous, exciting writing-based curriculum.

A well-founded worry is whether the pedgagogical theory and praxis elaborated here is too wildly aspirational to be workable in even the most advanced undergraduate classroom today. While certainly not dismissive of this vexing objection, *Citizen Science Fiction* nevertheless rejects the contention as beholden to the deficit or banking model of the educational enterprise that views students not as active and empowered citizens of the world but as error-prone and deficient victims of systemic educational failures. Indeed, it is extremely prudent as an instructor to always customize pedagogical ideas on a class-by-class, project-by-project, or lesson-by-lesson basis; it is therefore imperative to practice care and caution meeting students on their own terms through plentiful diagnostic and formative scaffolding, endeavoring always to stray well within Lev Vygotsky's "zone of proximal development" (Chaiklin 39)—that is, pitching the difficulty of an assignment significantly but not exponentially above a student's current level of reading, speaking, and writing proficiency. The unapologetically aspirational nature of teaching citizen science and sf literature and media in the writing classroom therefore elicits, applauds, and promotes deep, effortful, and sustained student engagement as not only possible but necessary both for optimal academic success and civic agency.

Citizen Science Fiction is an interdisciplinary intervention into literary history, cultural theory, and sf criticism, composition-and-rhetoric pedagogy, sf studies, and the philosophy and history of science. In particular, this book investigates the discourses and rhetoric of citizen science, and the mission to not only challenge the rising tide of scientific illiteracy and science disidentification but to deeply educate students vis-à-vis promoting a complex, sophisticated interrogation of scientific debates and discourse of public science. Making a critical contribution to the discussion of public participation in the often elite domain of technocratic science and its problematic intersection with commercial imperatives and political agendas, *Citizen Science Fiction* explores the present, history, and future of how a broadly conceived array of popular sf literature and media can serve not merely as a layperson or nonspecialist functionary to scientific regimes but can actively research questions, test hypotheses, assess risk, and help make scientifically informed policy. With a scattered few interesting exceptions, the chosen sf texts dramatize the problems and promises of broadening the scope of science into a democratized arena of citizenship with all the fantastic exaggeration and heightened gravitas of well-established sf tropes and icons—aliens, robots, clones, mind uploads, galactic empires—that eschews prosaically reflecting the mundane mechanics of citizen science in action, fascinating though they may be.

In addition to being aimed at working scholars and practitioners in the previously mentioned respective fields and specialties, the ideal readers of this book therefore also include a wider audience of professors, teachers, students, and general readers interested in a concatenation of sf, literature, science, and pedagogy. A primary organizational schema for the book is that each chapter revolves around recent developments in specific scientific disciplines, including biology, ecology, computer science, astronomy, and cognitive science. Drawing on a deliberately wide-ranging swath of sf literature and media, the book chapters then make the case that the sf genre can help rethink citizen science as a phenomenon that should be understood as significantly more than a simple amateur hobby or consumer preoccupation; rather the sf genre provides a notable precedent, aspirational trajectory, and affiliative kinship for viewing citizen science as a movement that can reshape the current state of scientific discourse and culture in progressive, diverse, and participatory directions.

Hence, alongside a critical intervention into existing scholarship on citizen science and writing pedagogy, the book also performs a close, sustained inquiry into a wide-ranging swath of sf literature and media with a view to its overlooked engagement with a citizen-science framework. Only *Teaching Science Fiction* (2011), edited by Andy Sawyer and Peter Wright, comes closest to the thematic focus of this book; however, that valuable anthology is a broad introduction to teaching sf literature by respected professionals in the sf field and does not directly engage with science studies, pedagogical theory, or the issues of participatory citizenship. The utility of deploying sf in writing pedagogy is an underrepresented topic in the publishing world more broadly, and although there are an increasing number of nonfiction mass-market works categorizable as citizen-science popularization—Caren Cooper's *Citizen Science* (2016) and Mary Ellen Hannibal's *Citizen Scientist* (2016) come freshest to mind—none have viewed the phenomenon precisely from the perspective of *Citizen Science Fiction*, which stresses the role of sf literature and writing pedagogy in both galvanizing and critically reflecting on the growing movement of citizen science. Some touchstone sf authors whose particular works are to be closely considered include classic masters such as Isaac Asimov and Arthur C. Clarke, New-Wave innovators such as Samuel Delany and Ursula K. Le Guin, and contemporary practitioners such as Cixin Liu, Nnedi Okorafor, and Kim Stanley Robinson.

I am a professional educator who studies both sf and composition at the university level and who regularly teaches these very same authors, texts, and discourses, so I speak in part from my own professional needs, background, and experiences when I note the conspicuous gap in scholarship on this particular constellation of concerns. The target audience for *Citizen Science Fiction* therefore consists of sf readers, scholars, researchers, and teachers for which there is a large and growing membership. The book also hopes to intervene in issues of popular science appealing to general readers as well as undergraduates and graduate students, and the book likewise hopes to appeal to those invested in discourses of citizenship, democratization, and public policy more ubiquitously. Finally, *Citizen Science Fiction* aims to interest citizen scientists themselves, welcoming them into dialogue with scholars and students of sf works of literature and media, as well as pedagogues and professional rhetors invested in problems of arguments, inquiry, and writing, especially those interested in exploring the protocols and rationale for incorporating into the composition classroom rich, exciting, and powerful sf readings in concert with the multimodal array of popular and entertaining sf music, film, television, comic books, and video games.

Chapter 1 discusses the pedagogical theory of threshold concepts as a tool for students to absorb and digest the biological gene as both a scientific idea and a rhetorical trope, intermingling life science and biology with their own personal narratives, educational goals, and social worlds.

Specifically, the chapter considers the citizen-science project *EteRNA* for its gamification of molecular biology that can serve as an initial transporting gateway for deeper and more extensive investigations by students of genetic discrimination, informed consent, consumer genomics, and gene editing. Closely reading sf writers as divergent as Robert Heinlein and Octavia Butler, and media from artists as wide-ranging as David Bowie and Denis Villeneuve, the chapter contends that critically interrogating fictional representations of genetic engineering for the sake of life extension, cloning, or other posthuman augmentations help students observe, parse, and dissect what is at stake for them in biological citizen science. The biopolitical coordination and management of genetic information threatens to erode the privacy, safety, and dignity of volunteer participants, and yet the very same participants can greatly benefit from medical advances achieved through the decentralized networks of citizen science and the open-source sharing of data. This critical focus on the increasingly important role DNA-literacy plays in culture as canvassed in the writing classroom promises to deliver on what the Human Genome Project ambitiously designated as the Ethical, Legal, and Social Implications (ELSI) of its myriad scientific breakthroughs, which have only begun to saturate and transform the everyday lives of students.

As a rejoinder to the apocalyptic tones of panic and doom that often swirl around discussions of climate change, chapter 2 offers some sober-minded thoughts and practical suggestions on the pedagogical implementation of the citizen-science project *iNaturalist* into the writing classroom. The chapter also draws on the growing critical literature on the concept of the Anthropocene, and its dynamic vision of the planet Earth as one singular complex system in the process of massive geological change induced by the human consumption of fossil fuels, to highlight for students the value of articulating uncertainty in writing given the widespread collapse of established binaries between the local and the global, the ancient and the contemporary, the public and the private, the humanistic and the scientific. The chapter suggests that sf literature by both classic and up-and-coming writers such as N. K. Jemisin, Frederik Pohl, Kim Stanley Robinson, Charlie Jane Anders, and Sam J. Miller, in conjunction with popular music artists like Bjork and filmmakers like Alex Garland, can be productively mined in the service of helping students navigate the postnormal complexity of sustainability issues and ecological science.

Chapter 3 investigates the intersection of artificially intelligent (AI) natural-language generators and citizen science, such as *Shelley, A.I.*, to help understand the value of the human voice in student writing. With the clear and present possibility that sophisticated algorithms will be able to pass the fabled Turing Test and simulate written communication—not to mention drive cars, perform surgeries, provide banking services, and police neighborhoods—pedagogy must adapt to guide students whose future jobs and vocational callings will often be marked by either competing with or analyzing, training, and supervising these automated programs. The chapter invokes John Passmore's rejection of "aristoscience" to argue that the apolitical authority of artificial intelligence needs to be challenged by deeply investigating, and perhaps recuperating, the mythology of voice in the writing classroom. The pervasive trope of AI computer programs in sf literature and media, as shown in media as disparate as Marvel superheroes and Star Trek androids, as well as prescient literary works by Arthur C. Clarke and Samuel Delany, among many others, frequently reveals the imperative for people to engage in civic-minded reflection, inquiry, and ingenuity that even the most versatile and powerful machine intelligence cannot yet emulate. Just as AI research increasingly and unconsciously replicates the ingrained, systemic injustice, inequalities, and prejudices of our social and political life, so the creation of the supple and robust techno-literate public sphere needs to be factored into our educational goals and pedagogical priorities of our fast-approaching plunging into the AI-dominated near future.

Chapter 4 concerns incorporating citizen science projects that incorporate cognitive science and neuroscience into the writing curriculum. Its primary pedagogical argument is that due attention paid to mind-set theory, cognitive literary criticism, and studies of emotional and social intelligence can overcome an inherent bias of educational curriculum toward overemphasizing abstract logic-driven rationalism and dispassionate inquiry in the teaching of writing. Since the New Wave transformations of the sf genre, plumbing the psychological recesses of cognitive inner space has been a mainstay of sf literature and media, as deployed through a variety of popular tropes, such as psionics, synthetic drugs, brain disorders, future histories, alien gods and monsters, parallel universes, and mind uploads. The cognitive-science rhetoric-and-composition pedagogy of this chapter expands on the citizen-science game *EyeWire*, and its contribution to the mapping of the connectome, the tangled network of neural wiring in the human brain, as well as other innovative neuroscientific projects. The scientific study of the connectome is an attempt to overcome the replication crisis of psychology and other social sciences, a legitimation struggle that a crowd-sourcing platform such as Amazon's Mechanical Turk is both remedying and exacerbating. The chapter analyzes works from writers as various as Philip K. Dick, Greg Egan, Gene Wolfe, and Justina Robson, as well as the music of Pink Floyd and the filmmaking of the Wachowski siblings, to explore the emotional and cognitive phenomena of empathy, creativity, resilience, madness, and embodiment.

Chapter 5 draws on recent waves of pedagogical research on the advantages of promoting a stance of contemplative mindfulness in the classroom, helping students adopt a calm, meditative presence of mind that paradoxically manifests an attitude amenable to both alert discipline and tranquil ease, both relaxed stillness and keen attention, when engaged in the writing process. The chapter contends that mindful student learning and success can be encouraged by consistent teacherly appeals to listening rhetoric, invitational argumentation, and nuanced civility discourse. Moreover, this final chapter argues that a recent data deluge in astronomical science has provided an embarrassment of scientific riches that a writing instructor can exploit in the service of promoting the mindful contemplation of the cosmos. Gleaning closely read examples from media and literature crafted in the popular sf subgenre known as "space opera," the chapter shows that space opera retools the splashy escapist tropes of galactic empires, alien contact, and faster-than-light starships for readers in a fashion that transcends merely an aesthetic or cognitive effect of sublime beauty and terror. Space operas, both as exemplified in the mass-cultural Star Wars, Star Trek, and Marvel franchises and fandoms, as well as in the hugely varied literary works of Nnedi Okorafor, Cixin Liu, Ursula K. Le Guin, and Greg Egan, represent current theories, perspectives, arguments, and discoveries of astronomical science that students likewise may contemplate through citizen science of Stardust@home, SETI@home, Galaxy Zoo, and the Zooniverse it spawned, which hosts projects devoted to observing galaxies, processing radio-astronomy data, exoplanet hunting, analyzing asteroid dust, and assorted other crowd-sourced astronomical undertakings.

The overriding question this book asks is a relatively straightforward one: Can the conjunction of citizen science, student writing, and sf literature and media, taken together as a heterodox combination, scale the seemingly insurmountable wall that is continually being erected between science and the public? The answer the book provides is, I hope, satisfyingly wide-ranging, but also unavoidably complicated, uncertain, mutable, and contingent. Yet a tantalizing promise can be tentatively offered. If boldly imagined by an energetic community of speculative writers, readers, students, and teachers, and bolstered by the ongoing transformation of scientific expertise and elite institutional structures, and, most importantly, buoyed by the enthusiasm and experience of a genuine grassroots groundswell of dedicated volunteers and participants, citizen science fiction does not have to be a hubristic contradiction in terms. Citizen science fiction, broadly conceived, can be a rabble-rousing call to arms among a camaraderie of citizens strenuously striving for a transformative change in how we understand, participate in, and reimagine citizenship, pedagogy, sf, and science and technology.

[illegible] both relaxed [illegible]

ONE

MaddAddamite Science

Teaching Citizen Science through the Threshold Concepts of Gene Culture

CITIZEN-SCIENCE–AFFILIATED PROJECTS, SERVICES, AND GAMES

Foldit and *EteRNA* Games
The Human Genome Diversity Project
New York Times Magazine "Diagnosis" Column, by Lisa Sanders
The Dog Aging Project
The Personal Genome Project
23andMe

SF LITERATURE AND READING

The Island of Doctor Moreau, by H. G. Wells
Brave New World, by Aldous Huxley
Methuselah's Children, by Robert Heinlein
The *Yesterday's Kin* Trilogy, by Nancy Kress
The *Xenogenesis* Trilogy, by Octavia Butler
"The Story of Your Life," by Ted Chiang
The *MaddAddam* Trilogy, by Margaret Atwood

SF MEDIA AND MULTIMODAL TEXTS

Alien, directed by Ridley Scott

Orphan Black, created by Graeme Manson and John Fawcett
The Rise and Fall of Ziggy Stardust, by David Bowie
One Beat, by Sleater-Kinney
Arrival, directed by Denis Villeneuve

In *The Island of Doctor Moreau* (1896), H. G. Wells's gentleman-scientist narrator, Edward Prendick, recoils in nihilistic horror over "blind fate, a vast pitiless mechanism" (96) after resorting to euthanizing an animal-human hybrid created by the titular vainglorious vivisectionist. The implosion of ethical-political bearings that Prendick has experienced on the forsaken island permanently alters his stalwart Victorian faith in a progressive Enlightenment worldview, consigning the world-weary narrator to live out the rest of his years in cosmic despair. Nevertheless, the narrative point of view that distances the literary author, H. G. Wells, and the fictional character, Prendick, is an important distinction to bear in mind. It is tempting to attribute the moral panic that frequently inflects the contemporary public reception of bioethics to a variant species of what Isaac Asimov, in an oft-quoted discussion on trends in sf in "The Machine and the Robot" (1978) famously called the "Frankenstein complex," namely, fearful commentary on the scientific overstepping of social boundaries stoked by those who had "no knowledge of science, no interest in it—and were in fact rather hostile to it" (443). Yet, given that such a putative Frankenstein complex clearly does not pertain to the lifelong passionate science advocacy of H. G. Wells, it important to recall that Asimov himself, in "The Science Fiction Breakthrough" (1979), stipulated to withhold critical censure from the Wellsian cautionary scientific romance: for Wells, as opposed to the rather dour Prendick, did not advocate that "it is better not to know" and instead urged "know but be careful how you apply" (184).

When teaching biological topics in the context of citizen science, tempering Asimov's embrace of pure science with Wells's caution toward applied technology when considering the ethical and social implications of modern biology can help unlock otherwise troublesome or forbidding concepts. The Frankenstein complex in this context may be understood as a subtype of what David Kirby calls "folk science," that is, a fictional or imaginary representation of science, which, even in the process of perpetuating certain misconceptions or mythologies, ultimately lends legitimacy to the scientific enterprise as a whole by crafting "images and narratives that convey the excitement of scientific research or communicate a sense of awe about the natural world" (117). Take, for instance, the citizen-science game *EteRNA*, a free online game created by the biochemist David Baker and the computer scientist Adrien Treuille, in conjunction with the universities of Carnegie Mellon, the Bill and Melinda Gates Foundation, and Stanford, in which thousands of crowd-sourced players manipulate nucleotides to assist in the laboratory synthesis of ribonucleic acid (RNA) molecules. A successor to the popular three-dimensional protein-folding game *Foldit*, *EteRNA* involves gamers in shaping molecules in such a way to optimize kilocalorie benchmarks by clicking on particolored circles, thereby transforming these shapes into the RNA bases of adenine, uracil, guanine, and cytosine, which in turn, according to well-known or newly discovered biochemical rules, fashion bonds that can manifest on larger scales into the intricate target designs of desired tails, loops, and rows.

Simply assigning students a writing task of journaling their experience playing *EteRNA* may serve as sufficiently instructive in itself, especially since such a lesson plan relies on principles of the "gamification" of pedagogy to supplement or substitute for the frequently disengaging transmission of content through a traditional chalk-and-talk or sage-on-the-stage lecture. Indeed, Kevin Bell in *Game On!* describes gamification as deploying "gameful elements" (3) in the classroom or lecture hall—through, for instance, displaying progress and leveraging competition in leaderboards, calibrating difficulty in the gradual acquisition of growing skill levels, providing discrepant attention-clock resets, and exploiting narrative buildups and payoffs. A key educational feature of the *EteRNA* initial tutorial game, for instance, entails the dramatic popping up of relevant scientific facts and theories that appear behind a spray of dopamine-inducing bubbles and gurgled sound effects once a player solves a brain-teasing puzzle. Yet a more integrative, scaffolded unit of material involving the higher-order evaluation of *EteRNA* as a tell-tale example of citizen science may ultimately prove most productive in a writing classroom, pedagogically speaking. Such collections of lesson plans permit students to investigate the life sciences under the domain of civic inquiry in a rich social and cultural context, namely, the public sphere of citizen science.

In a nuanced discussion of the personalized genetic testing service 23andMe, Barbara Prainsack devised a six-part schema for understanding to what extent a citizen-science project incorporates the involvement of citizen participants at a critically upstream or grassroots level. This schema asks questions about the following issues: coordination, participation, community, evaluation, openness, and entrepreneurship. A guided classroom deliberation on this schematic breakdown of citizen-science efficacy promises to help navigate for students, in both Wellsian and Asimovian ways, the polar extremes of both the Frankenstein complex of science phobia or illiteracy, on the one hand, and a naïve, celebratory scientism of endorsing citizen-science biomedicine marketing, on the other hand. Hence, in addition to Prainsack's questions, furnishing a class with some journalistic commentary on the game such as essayistic prose found in *Wired* magazine or in a popular nonfiction book like Jim Kozubek's *The Modern Prometheus* (2016) on the genomic science in question might initially broach issues regarding top-down coordination versus bottom-up openness of citizen science. As students will discover, *EteRNA* constitutes a complicated mixture of both unpaid, voluntary cognitive and affective labor—the research agendas are orchestrated by elite university scientists funded by wealthy organizations such as Google and the W. M. Keck Foundation—and shared communal benefits given that newly invented cures, vaccines, and treatments result from the joint discoveries and given that published articles include the byline of the complete list of *EteRNA* players as coauthors.

In *Mondo Nano,* scholar Colin Milburn argues that citizen-science games like *EteRNA* proudly proclaim their innovativeness for "upholding open-source approaches to intellectual property, even recognizing the contributions of individual gamers to potentially patentable inventions" (54). Likewise, student discussions will likely generate mixed and uncertain results in answering all of the complicated questions posed by Prainsack's inquiry. For instance, the vibrant gamer community of *EteRNA* that populates the forums, chatrooms, and wiki seems to be thriving. Indeed, in early 2011, as reported by Brendan Koerner in *Wired* magazine, after a series of initial missteps in which RNA molecules did not synthesize successfully in the labs, hundreds of players analyzed the feedback and data that they were supplied in order to improve their strategies. Ultimately, the *EteRNA* gamer community, through their profound engagement with the trial-and-error process of the scientific method, managed to not only fold the RNA molecules in the lab but also significantly outcompete the success rate of computer algorithms.

Nevertheless, the question of entrepreneurship and who profits from citizen science problematizes the altruistic and pedagogical rhetoric of these citizen-science games. The *EteRNA* Commons licensing agreement states that Stanford administers "donated funds and funds that come from *EteRNA*-related patents [and] become part of the *EteRNA* Commons fund, which can only be used to support *EteRNA*-related projects." Yet it is unclear whether such a general statement is coherent beyond its strictly legal scope or to what extent the various outcomes of the specific challenges that *EteRNA* sponsors fall under this umbrella of the *EteRNA* licensing agreement. For instance, in 2019, Stanford partnered with the Berkeley Innovative Genomics Institute for the challenge OpenCRISPR to assist gathering ninety-four thousand solutions for RNA molecule patterns that would be crucial in designing the activation and deactivation of gene editing in living cells. As spearheaded in part by Nobel laureate Jennifer Doudna at UC Berkeley, CRISPR (Clustered Regularly Interspaced Short Palindromic Repeats) concerns DNA (deoxyribonucleic acid) sequences that act as therapeutic scissors for the human "genome," or the approximately twenty-one thousand genes that serve as the instruction manuals of cell function in the human body. CRISPR can thereby locate and eliminate viral gene sequences via Cas (Crispr-associated) proteins, which entails more sophisticated gene-editing technologies than previously possible.

In recent court battles, though, medical applications of the discoveries pioneered by the CRISPR research of Berkeley have been separately patented by the Broad Institute of MIT and Harvard such that in 2017 the Food and Drug Administration (FDA) approved the first gene therapy drug for sale in the United States for the astronomical cost of $475,000 (Kozubek xxii) largely because of the influence of powerful pharmaceutical companies. This FDA-approved Chimeric Antigen Receptor Therapy (CAR-T) gene-editing technique, though, does not use the CRISPR gene-editing technology yet and holds life-saving promise for cancer patients who have proved resistant to other treatments. An important question that may then occur to critically questioning students and which may eventually lead to a fruitful, rigorous writing topic could be along the following lines: in a research-based argument essay, agree or disagree with the premise that even if the price of an advanced therapy such as CAR-T becomes significantly more affordable due to the strenuous efforts of researchers, doctors, advocates, and activists, the extreme neoliberal privatization of the tangible medical fruits of genomics threatens to undercut the otherwise meaningful participation of lay, uncredentialed citizens in biomedical citizen-science research.

In this manner, incorporating material on citizen science into the classroom becomes not merely a gamified instrument in an educational toolkit but a more fundamental means of recharging the critical interchange between the public and science. In their influential study on student learning, Jan Meyer and Ray Land, in the paper "Threshold Concepts and Troublesome Knowledge," conclude that it is pedagogically advantageous to structure lessons around "threshold concepts," or what in sf-inflected nomenclature one may call conceptual breakthroughs, paradigm shifts, or cognitive novae. In teaching pedagogy, threshold concepts are considered difficult ideas that initially seem counterintuitive, but, once absorbed and digested by students, thoroughly transform their understanding of a subject discipline. Arising from the deeper levels of thinking popularly analyzed in terms of Norman L. Webb's depth-of-knowledge pyramid or Benjamin Bloom's taxonomy of critical thinking, threshold concepts therefore make whole systems of learning stick with the student by promoting higher-order, more complex thinking about a subject as opposed to new material merely being named and ritually retrieved as inert pieces of discrete information.

Doug Downs and Liane Robertson apply threshold concepts, in addition to traditional student learning outcomes or product assessments, to general-education, first-year composition (FYC) classes to argue that writing as a discipline also entails knowledge barriers that students must overcome to expand their skills, aptitudes, and proficiencies: namely, the vital importance of the rhetorical situation, reading protocols, process writing, and other metacognitive tools of writing (117–18). Building on this composition research, it is important to recognize that these threshold concepts impact the educational experience of the student regardless of their scholarly areas of study or specialized academic majors, overlapping in particular with myriad discourses of citizenship informed by social forces and cultural rhetoric saturated with science and technology. Moreover, Charlotte Taylor expands on threshold concepts to encompass issues in teaching the life sciences and biomedicine directly by highlighting the importance for science-oriented teachers to extend the cognitive thread beyond demanding the rote recall of abstract biological processes—such as, for instance, the central dogma of molecular biology that sequences of nucleotides in genes flow to amino acids in proteins through DNA replication, RNA transcription, and protein translation.

Instead Taylor urges teachers who address the life sciences in their curricula to connect disjointed islands of knowledge to the more motivationally tangible terrain of concrete, real-world problems in the students' own lives, hence bridging "the threshold difference between students and scientists engaging in research who are creating their own challenges to extend frontiers of research" (92). The argument this chapter makes amplifies Taylor's contention to inform the reading protocols not only for budding undergraduate scientists in training but to a more wide-ranging grouping of student learners and beginning writers as well as more advanced readers and scholars endeavoring to meaningfully reframe and broadly understand the significance of scientific knowledge in culture and society today, with relevance for scholarship and pedagogy ranging from scientific disciplines proper to discourses from the humanities and social sciences such as Science and Technology Studies (STS), Cultural Studies, and the Philosophy of Science. For a student or reader less motivated by a healthy mixture of the Wellsian or Asimovian ethical or social implications of applied genetics in the world, for instance, there are still significant connections between what happens when a student internalizes the threshold concept of DNA as a biological abstraction and what philosophers Paul Griffiths and Karola Stotz explore as the epistemological "reductionism" of the unruly, historically idiosyncratic concept that has come to be associated with the term "gene," whether it is conceived as the still-relevant Mendelian unit of heredity or James Watson, Francis Crick, and Rosalind Franklin's seminal notion of the double-helixed informational coding sequence.

This book is certainly not alone in making one of its essential contentions that such science-oriented philosophical inquiry and ethical critique is not merely of incidental interest only to academic specialists or invested enthusiasts. More generally, Dominic DelliCarpini, for instance, shows that the relationship between composition and literature, or between reading and writing—notwithstanding the undeniable intimacy between certain genres, such as, in our case, those between popular science writing and sf literature and media—has always been equally fraught and inevitable, dating from a current resurgence of disciplinary debates back to John Dewey, to Sir Philip Sidney, and in ancient rhetoric to Cicero and Plato. Yet a wide variety of instructors, wishing to avoid an old-fashioned, belletristic, or abstractly literary approach to the writing process and tasked with teaching rigorous academic assignments, contend that literature, paired with the rhetoric of civic-mindedness, can more instrumentally serve the "pragmatic goal of inspiring deliberation and action" (DelliCarpini 30) vital to a democratic public. Such citizenship-oriented rhetoric and pedagogy challenges a naïvely narrow or overspecialized educational framework that compartmentalizes students into separate but equal silos of interest and engagement.

Hence an approach to teaching writing about contemporary biological sciences by means of citizen science entails teaching threshold concepts of the gene as an ideational category from its initial development as a hypothetical unit of heredity first proposed by a once obscure Austrian friar to what is viewed today by Big Science as a dense network of genetic triggers and switches that modify, interpret, and integrate the myriad features of human biology, not to mention the complex epigenetic web of environmental factors of culture, cognition, hormones, and social and cultural behavior that are now standardly considered crucial in the expression of genes. The gene as threshold concept promises such pedagogical benefits precisely because it allows students to renegotiate personal and social meanings through parsing their inherent ambiguities, surprises, and misreadings. As Siddhartha Mukherjee evocatively writes, "The genome is only a mirror for the breadth or narrowness of human imagination. It is Narcissus, reflected" (481).

LAZARUS LONG, BIOGERONTOLOGY, AND GENETIC DISCRIMINATION

A student's introduction to recent controversies over the social and ethical implications of the gene for citizen science can thus, oddly enough, be animated by acquainting students with the brash, outsized classics of pulp or paperback sf. From A. E. van Vogt's *World of Null-A* (1945) to Clifford Simak's *Way Station* (1963), early works of literary sf frequently addressed issues of immortality, rejuvenation, and longevity that now occupy the diverse resources of vast research institutions. One of the most popular pulp treatments thereof was Robert Heinlein's *Methuselah's Children*, first published in *Astounding Science Fiction* in 1941 but re-edited as a paperback novel in 1958. In his time, Heinlein became revered for elevating sf to credible mainstream status in part by acting as a cultural liaison for scientific debates and concerns; in "On the Writing of Speculative Fiction" (1947), Heinlein, for instance, set the ground rules for what would eventually solidify into genre sf by arguing that "in the speculative science fiction story, accepted science and established fact are extrapolated to produce a new situation, a new framework for human action" (219). Hence, originally predating the proper discovery of DNA by roughly a dozen years, *Methuselah's Children*, the first entry, by publishing date, in the Lazarus Long series, nonetheless registers and anticipates advances in genetic science that today receive continual heated debate.

In *Methuselah's Children*, the exceptional longevity of the protagonist, Lazarus Long, stems not from a speculative bioengineering coup, which only occurs at the end of the novel, but from an improbable cabal of legacy matchmakers who interbreed humans to live unusually protracted lifespans. Heinlein puts a genetic twist here on what Eric Rabkin details as a ubiquitous trope in global literature at large: namely, the hubristic search for immortality—the desire to be "changed into sterile vampires, childless angels, works of art, microchips" (xvi)—which undergirds works from *The Epic of Gilgamesh* to William Gibson's *Neuromancer* (1984). In his sf novel *The Fountains of Youth* (1999), clearly influenced by Heinlein, Brian Stableford has his scholarly mouthpiece character, Mortimer Gray, recount how the trope of the search for immortality recurs in non-Western literature as well, such as Tibetan mythology and culture (165–66). Likewise, in *Archaeologies of the Future*, Fredric Jameson contends that who is granted the social opportunity to prolong their human lifespans is a defining characteristic of the utopian drive and class consciousness of sf as a vast, variegated body of world literature (6). It must be noted, as well, that extending the glamour and vigor of youth indefinitely is also the refrain of the commercial nutrition and cosmetic industries whose lotions, drugs, and diets frequently lay themselves open to charges of quackery, if not outright fraud, according to their most vociferous critics. What makes Heinlein's novel particularly resonant today, though, for introducing students to these widespread biocultural discourses today is how the novel presciently anticipates the biogerontological quest for the aging gene that scientists are currently undertaking in their research regimes.

Biogerontology, the science of aging, takes inspiration from the study of animals since many animals—for example, sharks, alligators, flounders, and sturgeons—seem mostly not to die of old age per se but of disease, predation, or lack of resources. Citizen-science biogerontology can also exploit linkages between animals and humans to treat age-related conditions. Hence the Dog Aging Project, affiliated with the broader project of Darwin's Ark, asks pet owners to enroll their dogs to test and collect measurements on the genes and environments of these animals upon the administering of the drug Rapamycin. In the scientific paper announcing the citizen-science project, "The Biology of Aging," Matt Kaeberlein and colleagues at the University of Washington, declare the project wishes to target "the genetic modifiers of longevity, as well as specific cellular processes that degrade with age and likely contribute to age-associated pathology" (291–92). Such a study has obvious linkages with other studies in genetic biogerontology such as, for instance, those conducted by Christopher Wills, at the University of California, San Diego, who proposes to discover the aging gene—that is, the coupling of longevity and heredity, by means of genetic sequencing of aged people (Kaku 214). In the Dog Aging Project, the problem for the interested student that immediately arises vis-à-vis the concerned pet owner, however, is identifying to what extent does this project exploit citizen-science rhetoric to expand the pool of test subjects without technically violating scientific ethical guidelines. As Kaeberlein explains in an interview with *CBS News*, "The first question in my mind, when it came to clinical tests, was that I would never want to do studies on dogs in a laboratory environment. These are people's pets."

Given the archlibertarian tendencies of Heinlein, a cursory reader may be forgiven for the hasty assumption that in *Methuselah's Children* Heinlein intervenes in the early construction of such biogerontological ethical guidelines by privileging biological autonomy, self-determination, and citizenship rights as robustly guarded from intrusion to the point of sacrosanctity. H. Bruce Franklin argues, for instance, that in depicting the persecution of the wealthy Howard family and their elaborate masquerade that hides their secret longevity from the rest of the unknowing public, Heinlein not only values rampant biological individualism as paramount but unfairly hedges against any form of egalitarian and democratized project that may seek to capitalize on biological group resources for research: "We witness the citizens of the most humane, rational, libertarian, and scientifically advanced societies metamorphose into a ruthless, snarling horde of beasts, merely because other people have not obtained longevity" (40). Nevertheless, in a more charitable reading, Farah Mendlesohn notes that Heinlein's work often revolves around a central premise that "to contribute to society was, for Heinlein, a political act, an act of *civilitas*," (182) the ancient ideal of excellent public virtue and volunteerism. Indeed, Thomas Clareson and Joe Sanders contend that, due to the distribution of the longevity and rejuvenation treatments in the Lazarus Long series, eventually "the elite may constitute almost the entire population" (49). Grappling with a close reading of the novel may afford students a vivid illustration of a narrative embroiled in the irresolvable paradox between the top-down coordination and management of bioscience by expert authorities and institutions and the bottom-up, grassroots participation and engagement of a disseminated network of laypeople and nonspecialists. Citizen science regularly wrestles with this fundamental paradox between elite authority of scientific expertise and the disruptive incursion of communal values and norms of group solidarity.

In the novel, the Howard Foundation is a nonprofit organization formed after the U. S. Civil War and dedicated to the eugenic goal of increasing longevity among the offspring of its charter members. Instead of becoming a Socially Darwinist ruling elite, by 2075, the Howard Foundation family has been forced underground by fear of persecution, instituting a masquerade of forged identities until such time that the majority of future society can ensure the "increased respect for the custom of privacy and for the dignity of the individual" (9). By the time of the novel's setting, in 2136, due to rapid social and technological progress, the masquerade has ended only to arise in its place the long-dreaded resentment harbored by the short-lived against the so-called "Methuselahs." When threatened with mass torture and imprisonment, the oldest member of the Howard Family, Lazarus Long, proposes that the long-lifers flee the solar system in a stolen interstellar spaceship to escape genocide. Following an emergency democratic committee of the Howard Foundation, the picaresque rogue explains the philosophy of life for an artful dodger: "These things *pass*. Wars and depressions and Prophets and Covenants—they pass. The trick is to stay alive through them" (19). In articulating an ethos of biocultural anonymity in a desperate bid to escape the depredations of a predatory invasion of privacy, early Heinlein speculates on a bioethical framework that would protect against what later would become codified, by the Ethical, Legal, and Social Implications (ELSI) program of the Human Genome Project, as "genetic discrimination."

Genetic discrimination entails a violation of the responsibility to protect confidentiality of personal genetic data through the legal category of informed consent given that, if mishandled, such sensitive data can result in myriad risks for the individual: the denial of life insurance; the jeopardizing of employment; unreasonable search and seizure in criminal cases; discovery of nonpaternity; false positives or misinterpretations in medical diagnoses that lead to unnecessary surgical interventions or costly, dangerous treatments; the inequitable patenting of genetic data; or the anxiety and depression that may accompany the revealing of serious medical conditions that cannot be effectively treated. Interestingly, an offshoot of the Human Genome Project—namely, the Harvard Personal Genome Project, an ongoing citizen-science experiment in which participants submit blood and saliva samples to have their genome sequenced and shared in open-source databases online—attempts to temper the ethical implications of genetic testing by celebrating the *sui generis* civic service of its participants. In the study guide for the full-consent form of the Harvard Personal Genome Project, Madeleine Ball states that the decision to forgo anonymity by the participants must stem exclusively for a beneficent regard for the public good: "The public release of genome data involves numerous risks and is unlikely to provide any immediate benefit to participants" (3). Investigating with students as to whether they consider participation in the Harvard Personal Genome Project noble or reckless, admirable or ill-advised, or more likely, some dynamically heterogenous mixture of all these sentiments, may lead to productive, rewarding class discussion on a complicated bioethical quandary.

A definitive product of its own time, *Methuselah's Children* struggles with the bioethical implications of the eugenics and Social Darwinist movements that were institutionalized into the reigning scientific establishments of their era and responsible for repeated waves of brutal sterilizations that swept the globe since the burgeoning science of evolution was first popularized in the nineteenth century. Heinlein complicates the discriminatory biological essentialism of such eugenic thinking with an emphasis on the Howard Foundation not only becoming a persecuted minority but also perpetrating their own genetic discrimination by labeling some of their own members as "defectives" (29) generated by their inbreeding regimes. Nevertheless, the recourse to such reprehensible systems of eugenic classification in describing people as biologically defective in the first place, of course, presumes investment in popular pseudo-scientific discourses that have now been debunked among academic circles. However, culturally and politically, strong cases have been made that eugenics still reigns supreme in many sectors of society today, especially resurgent again through prenatal testing and consumer-oriented gene counseling and therapy, or the correction of genetic problems by introducing healthy genes to a person or embryo.

Richard Hofstadter shows how in its heyday some more nuanced strains of Social Darwinist rhetoric could subtly point out that "all you could say for the physically fittest survivor was that he managed to live" (83), while still dubiously asserting that capitalist competition mirrored natural struggle for biological dominance and thereby accelerated socio-economic progress. It is this strenuous libertarian streak in Heinlein's writing that has motivated some critics, such as Philip E. Smith to reasonably contend that "since the beginning of his career as a writer, Heinlein has based his most important fictions on a very apparent philosophy: Social Darwinism" (137). Such eugenics and Social Darwinist discourse recurs explicitly in Heinlein's later novel, *Time Enough for Love* (1973), which excerpts *The Notebooks of Lazarus Long* to aphoristically quip about the ethics of pruning the social organism prior to, as opposed to after, birth, given that Lazarus considers himself "an incurable sentimentalist" (246). Far from one of Heinlein's personal eccentricities, this position largely mirrors the official rationale of Oliver Wendell Holmes's U.S. Supreme Court ruling opinion that legally permitted enforced sterilization in *Buck v. Bell* (1927).

Methuselah's Children can thus serve as a culturally representative teaching touchstone because, despite such high-handed endorsements of top-down eugenic coordination, Heinlein also paradoxically privileges the participatory evasion of the biopolitical regulation of the individual. In subtle contrast to the Foucauldian notion of the self-policing of the individual through the sway of biological discourse, Sheila Jasanoff argues that citizen science can create a "civic epistemology" that contests the unidirectional deficit model that often accompanies scientific outreach and the public understanding of science; Jasanoff argues that constructing the citizen as a competent coproducer of scientific expertise is a viable alternative to genetic discrimination. In such a formulation, the citizen becomes "one who struggles with ambivalence in the face of competing cognitive and social pressures, robustly copes with ignorance and uncertainty as well as 'the facts,' and reserves the right to make moral choices about the purposes and governance of technology" (Jasanoff 254).

Likewise, for many readers, Heinlein's early fiction practically trademarked the superior sf citizen-hero for the genre: an unflappable, omnicompetent individual, or what the critic Damon Knight called "the Man Who Knows How" (83), that could function as pragmatic resistance to creeping despotism. Hence in *Methuselah's Children,* when the future interplanetary society known as the Covenant invokes the protection of "civil rights enjoyed by decent citizens" (58) to hypocritically suspend the "civil guarantees" (36) of the Howard Foundation due to the presumed withholding of a miracle longevity treatment, Lazarus Long, serving as the de facto leader or speaker trustee due to his immense age, deftly outmaneuvers the Covenant in the daring escape of the approximately ten thousand long-lifers. Once the diasporic long-lifers find a habitable planet on a faraway solar system, they encounter a species called the Jockaira, a scientifically advanced, collectivist species who lack the concept of privacy and whose architecture therefore lacks separate rooms (109).

Chafing at the prospect of having his biopolitical autonomy again disciplined and regulated by such an intrusive social regime, Lazarus Long's immediate suspicions are confirmed when the long-lifers discover that the Jockaira are being kept as domesticated pets by an even more advanced alien species (123). Similarly, after they resume their cosmic voyaging and the Howard Foundation happen upon another habitable planet, Lazarus Long characteristically refuses to conform to the social norms of its psychic inhabitants, the so-called "Little People," who have mastered a kind of genetic engineering (134) to such an extent that they exist in a utopian land of peace and plenty. Unlike his friend Mary Sperling who happily assimilates into the blissful collective fold of the Little People, Lazarus challenges the idea that the human body can be genetically "improved" (141) and still remain desirably human. Ironically, after returning home, the remaining members of Howard Foundation discover that the Covenant back on Earth has independently invented longevity treatments (150), thus reconfirming the value of the gradual scientific process galvanized by the civic pressures of knowledge-making, as opposed to science imposed by fiat from on high. The novel ends with a swaggeringly defiant Lazarus Long declaring his undaunted intention to eventually return to space in hundreds of years and belligerently confront advanced alien species on an egalitarian footing. Heinlein's ostensibly immortal citizen-hero strikes an aggressively imperialist and masculinist pose that nevertheless suggests the unrestrained agency of an informed and engaged public to bend even a science as sophisticated as genetic engineering to its own priorities.

DOLLY THE SHEEP, CLONE CLUB, *YESTERDAY'S KIN*, AND INFORMED CONSENT

The gene works as such an overarching transformative threshold concept for fostering discussion and writing in part because it speaks expressly to the pervasive contemporary milieu that attends the most sensationalized biological breakthroughs. For example, in 1997, Dr. Ian Wilmut's team at the Roslin Institute's laboratory in Scotland electrified the egg cell of a sheep in which the nucleus of an adult's mammary cell had been transferred; after 277 attempts, the resulting differentiating embryo was then implanted in a maternal surrogate sheep. Designed originally as a reliable delivery mechanism of nutrient-enhanced milk from genetically modified sheep, this otherwise mundane and pragmatic biotechnological invention sent taboo-busting shock waves of media attention around the world since the resulting sheep—cheekily named Dolly after the high-femme country-western singer Dolly Parton—could be considered an artificially created, genetically identical clone of her adult parent, even though Dolly retained separate mitochondrial DNA from her original egg cell.

As a cultural icon, Dolly became a "Rorschach test" (Nelkin and Lindee 83) that radically challenged the widespread scientific worldview perhaps most tellingly articulated by Mustapha Mond in Aldous Huxley's *Brave New World* (1932). For in that prototypical clone narrative, Mond hypocritically indemnifies himself against the ethical transgressions of genetically modified hatcheries of clones by claiming that such autocratic biotechnology remains indubitably in line with the disinterested, neutral pursuit of scientific-materialist truth about the objective world, while secretly censoring scientific conclusions inconvenient to the dystopian World-State (Huxley 162). In the sf imaginary, the cloning of humans frequently carries such disturbing connotations of rationalized, unchecked tampering with an individual's biological self-determination, if not outright sinister and conspiratorial overtones of technoscientific exploitation and eugenic solipsism, coupled with broader anxieties over genetic engineering more generally, as typified from works as varied as Gene Wolfe's *The Fifth Head of Cerberus* (1972), Ira Levin's *Boys from Brazil* (1978), or Kazuo Ishiguro's *Never Let Me Go* (2005). However, oppositely, some thoughtful and original exemplars of the trope, such as Ursula K. Le Guin's "Nine Lives" (1969) and Kate Wilhem's *Where Late Sweet Bird*

Sang (1976), also evoke a rich sense of mutual empowerment, interdependence, and sorority buoyed up by resistant cloned communities, or what David Higgins, by way of Eva Cherniavsky, terms "neocitizenship" (393) given the function of collective citizenship discourse as a strategic constraint on the biopolitics of neoliberalism.

Recently, the television show *Orphan Black* (2013–2017) has retooled the clone trope in a densely layered and nuanced fashion that may be commended to students as supplemental or primary material for a classroom seeking to understand the cultural and social context of specialized biotechnological coups, as typified in popular media by the cloning of Dolly. A wide-ranging spectrum of responses followed swiftly on the heels of the Dolly phenomenon, from the spawning of libertarian activists and advocates, such as Randolfe Wicker, the founder of the Cloning Rights United Front, who borrowed political rhetoric from the gay and lesbian rights movements to struggle against the cloning bans sweeping the world, to the more meteoric rise of socially conservative critics who fiercely condemned cloning as morally repugnant to traditional family values, such as Leon Kass, the chairman of President George W. Bush's Council on Bioethics from 2001–2005. Likewise, in *Orphan Black*, a secret cloning experiment in the 1970s called Project Leda ushers in not only hundreds of clones but a byzantine array of fringe-dwelling subcultures, from the bio-hacking Neolutionists affiliated with the elite corporate entity known as the Dyad Institute who own interests in the agricultural and pharmaceutical industries, to the Proletheans, a rural fundamentalist religious cult, with obscure ties to another paramilitary organization associated with a cloning experiment called Project Castor. The Leda clones at the center of the narrative of the show—all remarkably played by the same versatile actor Tatiana Maslany—similarly emerge from a variegated mosaic of social and cultural backgrounds: Sarah Manning, the titular orphan black, who is smuggled away as a child from the surveillance of her Dyad monitors, becomes an underworld con artist; Helena, a Prolethean-trained, mentally unstable assassin; Cosima Niehaus, a biology graduate student dying from cancer who eventually works for the Dyad Institute under the Neolutionist Dr. Aldous Leekie; Alison Hendrix, a suburban soccer mom; and Rachel Duncan, an amoral high-powered executive at the Dyad Institute.

In the second season finale, "By Means Which Have Not Been Tried," Sarah, Alison, Helena, and Cosima celebrate Sarah and her daughter Kira's escape from being hunted, surveilled, and kidnapped by Rachel and the Dyad Institute by spontaneously dancing together in Sarah's foster brother Felix's apartment to reggae-funk music. The kinetic clash of dance moves, outfits, and hairstyles enmeshes these genetically modified bodies into what Felix calls a Clone Club, whose temporary moments of jouissance, relief, and solidarity vividly represent what the dialogue of the next scene makes explicit. Resting in the same bed, Cosima, breathing tube hooked to her nose because of a cancerous side effect of the cloning experiments, explains to a meditative Sarah that the segmented horn icon of her wrist tattoo symbolizes the Golden Ratio. Geeking out, Cosima rehearses a Buckminster Fuller routine, marveling over the underlying mathematical patterns in natural phenomena as diverse as honeybees, flower petals, galaxies, and genes. The streetwise Sarah, bemused by such scientific epiphanies, states, "We're so different, all of us." This elastic push-and-pull tension between underlying scientific sameness and radical cultural difference among these socially divergent but genetically identical clones echoes the earlier repetition and variations in the clone-club dance. Playful, incoherent gyrations of participatory (neo)citizenship intermesh with the synchronized rhythms of the putatively rational, orderly, and coherent constructions of bioscience. A lively conversation among students over the sf meanings of this resonant scene can therefore transition imperceptibly into a rigorous, analytic writing assignment on the active, citizen-oriented construction of genetic science in our own technologically sophisticated, media-saturated lives.

The cloning of Dolly, for all the often overblown playing-god media controversy that ensued, did manage to reignite the bioethical debate over the efficacy of informed consent as a biomedical principle. As Sheila Jasanoff argues, although informed consent presumes the untrammeled volition of a patient, and legally dates not only to the Hippocratic Oath but also to the Nuremberg Code that followed on the heels of genocidal Nazi experimentation by the likes of Josef Mengele, not to mention the legal codification of the research ethics of Internal Review Boards that responded to the U.S.-based Tuskegee syphilis and Milgram shock experiments, informed consent also reinforces "lay-expert divisions" because "the very idea of beneficence set[s] up a seemingly unbreachable hierarchy between the caring expert and the cared-for patient" (177). In line with her scrupulous rejection of a libertarian championing of a "simplistically individualist idea of scientific progress" (Franko 137) in her Hugo-winning *Beggars* trilogy, Nancy Kress, in the trilogy that begins with the Nebula-winning *Yesterday's Kin* (2014) interrogates the putative attitude of altruistic caring that bolsters an elite class of technocratic biomedical expertise. Such monopolization of what constitutes information, cordoned off as inaccessible by gatekeeping medical professionals, researchers, and experts, overlaps with one of the most troublesome aspects of assimilating threshold concepts, such as the cultural and scientific significance of the gene, in a pedagogical setting—namely, the difficulty of absorbing isolated pieces of expert disciplinary knowledge that may seem alienating, meaningless, random, or ritualistic to the neophyte student.

Echoing such complicated journeys back and forth through the portals of scientific discovery, *Yesterday's Kin* follows the evolutionary geneticist Marianne Jenner whose stumbling onto a new haplogroup of mitochondrial DNA suddenly makes her a high-profile advisor of an elite, world-changing scientific project, as an alien species lands on Earth and constructs a floating pavilion dubbed *The Embassy*. Hailing from an orange-dwarf star called Deneb, these peaceful, communal aliens, genetically nonterritorial and nonhierarchical, are strange vicariants of humanity and carriers of the same DNA markers that Marianne Jenner has coincidentally discovered; for it appears that the Deneb, thin, deep-chested humanoids with large eyes, had been mysteriously transported from Earth to the alien planet they simply call World approximately 150,000 years ago.

Putatively more advanced in physical sciences yet essentially coeval in the biological ones, the Deneb exploit the legitimacy granted science by the public and governmental institutions and corporate industries of Earth, manipulating the implicit trust afforded the scientific-military establishment by manufacturing an emergency of an incoming spore cloud—technically composed of genetically linked viruses—that has drifted toward the Milky Way galaxy. Refusing to leave *The Embassy,* the Deneb strategically shroud their visit in secrecy, since only they know that humanity has previously evolved an immunity to the spore cloud. For the Deneb desire blood, saliva, and tissue samples from humanity to save their own species for whom the spore cloud portends extinction back on World. When the Deneb Ambassador Smith reveals the elaborate ruse after *The Embassy* has transfigured back into a starship for the return home, the justification relies on a mercenary logic of mass deception: they did not tell the complete truth because otherwise humanity would not have "allocated so many resources, so much scientific talent, or such urgency into the work" (186). In exchange for this nonconsensual genetic data gathering, the Deneb gift the engineering specifications for a star drive, yet the manipulative violation of informed consent cannot but leave Marianne feeling tragically cheated, especially since the concocted emergency has caused on Earth "riots, diversion of resources, deaths, panic, fear" (187). Moreover, Marianne's own adopted son, the drug-addicted black sheep of the family, Noah, has voluntarily joined the Deneb in their return home, given that the Deneb, who nurture strong kinship ties, have recruited and genetically modified select humans from their own rare genetic haplogroup. When a terrorist organization ineffectually succeeds in exploding a suicide bomb on *The Embassy* before all the Deneb abscond, a victim of the needless attack, Marianne, nevertheless channels the furious rage of the isolationist terrorists at being violated this egregiously. This rage bristles over the discursive boundary between the caring, beneficent medical informer and the free, putatively uncoerced informed subject—that is, the elite alien domain of clannish future science and a general rung of powerless humanity conceived as a vast disposable clinical resource and ultimately discarded like so many baffled lab rats.

Nancy Kress rings the changes on this thematic conflict between scientific authority and citizen-oriented dissidence in the sequel *If Tomorrow Comes* (2018). In the realm of real-world citizen science, such resistance can be readily seen in the autodidact tendency of patients becoming their most forceful advocates, as evidenced, for instance, by Steven Epstein in the noncompliance of activist AIDS patients in life-threatening placebo-controlled trials. In an important work of Science and Technology Studies, Epstein argues that the doctor-patient relationship has shifted in recent years such that "doctors are increasingly inclined to acknowledge the full subjectivity of their patients" (206). This cultural shift away from viewing patients as passive, poorly informed hosts of bodily illness has only accelerated with the rise of web-driven crowdsourcing, as the medical diagnostician Lisa Sanders illustrates in her *New York Times Magazine* columns that since 2010 have regularly solicited community-based research for particularly intractable medical enigmas: "In medicine, doctors accept that no one knows everything. . . . The Internet offers the possibility of a broader community—a sea of strangers linked by our medical curiosity and by our keyboards" (Sanders 28). Likewise, Kress's sequel inverts the established lay-expert hierarchy when, after humanity builds a starship based on the Deneb's gifted science and travels to World, also called Kindred, with Marianne Jenner on board, they discover that the Deneb have not in fact created a vaccine for the spore cloud and face imminent extinction. Even the advanced life sciences have been invented autonomously but only mysteriously discovered in the recent excavation of ancient titanium tablets. Now the technologically underdeveloped Deneb must rely on Marianne Jenner to synthesize a vaccine in time or become witness to the demise of their entire civilization.

The demotion of the Deneb from marvelously credentialed to practically nonscientist and Luddite, and the corresponding reversal of humanity from fringe self-advocates to epistemic knowledge-makers, emulates the reversal endemic to knowledge acquisition in the learning process. Once a threshold concept is acquired, that novel conceptual anchor leads the student not only to the integration of a new discursive terrain but to newfound educational power dynamics as well. The oscillation through a liminal portal of understanding means not only that students begin instrumentally thinking like budding professionals but also that students draw on their own prior experience to develop the double vantage of defending their customary, not yet discarded cultural concepts in question as well. *If Tomorrow Comes* makes it clear that the Deneb, and their home planet of Kindred, as avatars of alien biomedical citizen science, can be seen from such a double vantage as well; for in addition to temporarily wielding their mysteriously endowed, immense technoscientific mastery, they have also democratically chosen as a matriarchal species to resist unchecked technological-capitalist progress so as to avoid "exploit[ing] resources beyond sustainability" (97). Therefore, their civilizational halt represents a resistant, civic-minded alternative to excessive deference to scientific credibility and technocratic authority. Kress stresses this critique as the novel begins not only with the human crew reeling from the discovery that the interstellar trip has led to an unexpected time dilation of twenty-eight years, but that Russia, which experienced a genetic anomaly in which thirty percent of their population died due to their spores, has wrought devastating revenge for perceived wrongs on an unsuspecting Kindred by building a warship to bomb Kindred's major cities.

In a surprising twist, it turns out that the natives of Kindred, in fact, cultivate a sublime disinterest in scientific curiosity that may tangibly represent for the citizen-science reader an exemplary case of the manifold risks of such a public stance of posttruth disregard and antipathy. The Deneb denizens, for instance, do not investigate the radio signals emanating from the original starship that voyaged to Earth as it orbits Kindred. Disaffected herself from the ideological idea of pure or neutral science, Marianne muses that the Deneb consider the decarded technology of the orbiting starship to be somehow spiritually "unclean" and therefore approximately equivalent to the various Terran concepts of "trief, marime, desecrated, haram" (195). In an underground bunker called Haven—which secretly engineers transistors and breeds terrorism against the human visitors—a radical insurgency of Deneb have discovered that the radio signals indicate that the local mouselike animals, leelees, have evolved a virophage that might, if released through aerosol, resist the spore cloud. Marianne's genetic-engineer background primes her to realize that virophages contain genetic transfers of ancient host DNA such that the vaccine can be targeted. With the assistance of another dedicated human scientist, Branch Carter, Marianne has a sudden realization that the code to initiate a call-back device and obtain the virophage from the orbiting starship is the human genetic sequence itself (298). Extinction can only be warded off provided that the gene as an informational key has been internalized, a threshold concept that the deliberately unadvanced Kindred are perfectly resigned to forgo to their detriment. Once the vaccine is released, only one twenty-fifth of the Kindred population dies, thus saving millions of lives and constituting a qualified "scientific triumph" (322).

In the final installment in the trilogy, *Terran Tomorrow* (2018), Kress again both legitimizes and critically examines the established credibility of technoscience, suggesting that meaningful participation in scientific enterprises requires continual questioning as well as vested engagement. The returning crew of *Friendship* discover that Earth has experienced a practical extinction vortex in which 96 percent of humans die given that radical climate change scientist-activists, the Gaiists, genetically modify the spore virus into an apocalyptic bird-borne pandemic. In a bid to create another vaccine, Marianne's grandson, Colonel Jason Jenner, nevertheless, rescues some of the remaining scientists and their surviving families and relocates them under a protective dome, manufactured by reverse engineering alien technology, in Monterey, California. Moreover, after returning to this ruined home, Marianne and other recipients of the original virophage vaccine enter a mysterious, microbe-induced coma that rewires their brains, greatly enhancing their functional intelligence.

Once awakened, the scientist Toni Steffens is able to devise a gene drive via genetically modified histone proteins to render the disease-carrying sparrows infertile; although this bold proposal may provide a chance for humanity to survive in the long run, untrained nonscientists are appalled by the sheer hubris and naiveté with which idealistic science-driven ideologues implement this plan without consultation or oversight: "Inserting this gene drive into more birds and then releasing them into the wild was a political decision, with enormous ecological implications" (293). Despite this unnerving uncertainty, the series ends on a soberly optimistic note, as the awakened superintelligent board *Friendship* to return to Kindred, and thereby deliberately prevent their spore-activated technoscientific prowess from spreading to the militaristic factions on Earth, such as the warmongering ethno-patriotic group called New America. With her augmented, virophaged insight, Marianne wonders whether the alien superbeings who created the starships, dome shields, and instigated the hunt for spore vaccines, were in fact time-traveling future humans with the hyperadvanced proficiency to genetically engineer and seed ancient microbes. This microbe would have triggered a Second Great Leap Forward "that made the first one, seventy thousand years ago, look like a not-very-bright child triumphant at piling just one block on another" (329). This ambiguous ending demands rich and nuanced classroom evaluation, back-and-forth dialogue, and conversational hashing out. On the one hand, Kress frames official genetic-engineering research as on a hegemonic foreordained trajectory toward enlightened progress; and yet, on the other hand, Kress undermines the problematic efforts of such scientific networks of actors, devices, and interests to be the improbable result of bootstrapping so marvelously futuristic it might as well be alien.

ALIEN CHIC, THE HUMAN GENOME DIVERSITY PROJECT, AND BIOSOVEREIGNTY

The lack of a definite article in the title to the film *Alien* (1979) vividly illustrates the deep-seated condition of alienation, as opposed to a merely tantalizing encounter with an exotic entity. Such alienation couples pedagogically as well with genetic threshold concepts, which civic-minded students may experience when seeking to understand the scientific, philosophical, and cultural implications of contemporary discourse circulating around biology. As opposed to the print work of Hal Clement, Jack Chalker, or James White who cleave to more rigorous extrapolations of xenobiology in their hard-sf imaginings, commercial cinema is habitually prone to depicting the "Hollywood alien," a so-called Bug-Eyed Monster (BEM) whose aura of costumed menace typically emanates from its indiscriminate homicidal mania or its inexplicable desire for interspecies coupling. Making exceptions for the surprising coincidences of convergent evolution, Clifford Pickover nevertheless articulates the commonly observed point that, from a strictly xenobiological perspective, Hollywood aliens are frequently "too human-looking, considering the quite different evolutionary pathways we'd expect on different worlds" (15). In contrast, *Alien* explores a much more striking example of physiological alterity, or alien otherness, in the creature design of the Xenomorph, adapted for the screen by the gothic artist H. R. Giger himself from his Lovecraftian lithograph *Necronom IV*. As much as there are obvious humanoid parallels in this uncanny alien design, such as the bilateral symmetry, the enlarged cranium, the jointed legs, the digital appendages, and the protruding rib cage and spine, there are many biomechanical elements that are more strangely reptilian, insectoid, and ichthyic, such as the armored exoskeleton, the tail stinger, and the weaponized pharyngeal jaws. Combined with its nonhuman life-cycle—from egg, to larval Face-Hugger, to incubating Chest-Burster, to the sleekly mature Xenomorph—the nonhuman physiology of the alien illustrates the conceptually difficult evolutionary principle that genetic mutation is randomized and permutational, not goal-directed and linear.

Unsettling on an existential level, the cinematography of *Alien* lingers on the elegance of this randomly fittest Darwinian alien design, which seems to hold an intoxicating allure for characters endowed with established scientific clout and respectable technical expertise. In this gritty, dark future, the Executive Officer Kane (John Hurt) serves as a representative of the rapacious transnational Weyland-Yutani Corporation that decidedly seeks not to champion the humanistic, curiosity-driven spirit of pure cosmic exploration but rather to ruthlessly extend its greedy capitalist and neocolonial interests into the exploitable final frontier. Volunteering to boldly investigate a derelict alien ship, Kane's Beckettian credo of weary space exploration—he mutters, "We've come this far. We must go on. We have to go on"—leads to his shockingly grisly demise as the alien's first victim. Likewise, the android Ash (Ian Holm), an avatar of technological supremacy wielded by the Company from galactically afar, obscenely fawns over the alien's value-neutral transcendence of ethical and political constraints. Narcissistically projecting his own self-regard, Ash muses, "I admire its purity. A survivor . . . unclouded by conscience, remorse, or delusions of morality." Oppositely, commandeering the *Nostromo*'s computer interface, the protagonist Ripley (Sigourney Weaver), an otherwise subordinate Warrant Officer, manages to eventually resist the sinister Company's plot to commodify the alien at the expense of an "expendable" proletarian crew. The untrammeled autonomy of credentialed and qualified scientific authority reinforces its relentless pursuit of profit-driven biological knowledge only to increase its own transnational capital, but Ripley, who must outsmart both the alien and the technoscientific company to escape and survive, ultimately subverts the otherwise overweening protocols of such elite expertise.

Comparing the alien to the endoparasitic Ichneumonidae wasp that shook Charles Darwin's Victorian faith in a benevolent deity, Roger Luckhurst contends that the sociobiological debate over the "implacable arms race" of species competition in Darwinian discourse derives in part from "the neoliberal times that were beginning to emerge out of the crisis of the 1970s" (*Alien* 57). Additionally, it is useful to draw attention to the acid blood that spurts when cut during the attempt to remove the Face-Hugger, which the blue-collar spacer Parker describes as a "wonderful defense mechanism"; this acid eats evocatively through successive deck floors and ceilings and threatens to burn through the hull of *The Nostromo* if the alien were to be recklessly killed. For students engaging in a dialogue with this film franchise in the twenty-first century, the Darwinian, scientifically materialist toxicity of this horrific, near-invincible, and supremely adapted monster in a haunted (spaceship) house should be taken as both a literal narrative threat and evocative of a more broadly troubling speculation about the underlying concept of biological evolution itself. In *Darwin's Dangerous Idea,* Daniel Dennett, for instance, argues that the "universal acid" (63) of Darwinian selection and adaptation debunks misconceptions of specialness, teleology, and imposed orders historically assigned to speciation events by theological and humanistic society and culture. Thus wrestling with the very randomness, noise, and imperfection of radically different alien physiology, students may also address in directed essays the implications of conceiving genetic processes as mindless algorithms that indirectly design possible organisms simply to survive, reproduce, and transmit their genetic material in the cosmic environment. In contrast to Dennett, an alternative approach that, as Carl G. Herndl demonstrates, may indeed lend itself to equally fruitful discussion in a writing classroom is the scientific and cultural work performed by Stephen J. Gould's liberal-humanist inveighing against E. O. Wilson's sociobiology of the gene and the corresponding critique of the so-called "adaptationist" paradigm, as opposed to theories of random mutation or genetic drift, prevalent in evolutionary biology.

Some sf narratives of alien encounters—as evidenced by the neo-Darwinian enigma of inimical hostility of *Alien* or the impenetrable mystery planet of Stanislaw Lem's *Solaris* (1961)—rest on the reasonable assumption of the untranslatable difference of alien speciation. In contrast, Ted Chiang's short story "The Story of Your Life" (1998) rests on the provocative opposite assumption of the *over*-translatable nature of alien communication. The rigorous thought experiment of Chiang's story therefore emulates approaching the threshold concept of the gene not from the misconceptions of prior knowledge about biological difference but from the newly acquired perspective of integrating genetic theory into a significant transformative framework for understanding cutting-edge advances in the biological sciences. Moreover, in the curriculum of a composition class, conceptualizing the vagaries of alien communication in all the physiological complexity and nuance of this story precisely mirrors the rhetorical and linguistic context of the provisional writing process itself. The fundamental premise of the story is that nine looking-glass devices appear on the planet and stir a media-saturated sensation. The aliens are squidlike heptapods who, due in part to the radial symmetry of their bodies, have developed a nonlinear system of orthographic language; once this written language is painstakingly learned by the linguist Louise Banks, due to an extreme application of the Sapir-Whorf hypothesis, it reshapes her cognition such that time itself is understood as a nonlinear, foreknown simultaneity.

Passages of the story are deliberately confusing on a grammatical level; they are written in the second person and future tense, as Louise addresses her daughter and anticipates her daughter's experience growing up and tragic young death with uncanny, seemingly retrospective detail. Such narration makes the story a profoundly moving meditation on grief, memory, contingency, and free will, all realized as a logical extension of an encounter with an alien's radically different, evolutionarily estranged, but essentially similar genetic makeup. During a perfunctory gift exchange, Louise laments, "Truthfully, I wished the heptapods had given another xenobiology lecture, as they had on two previous exchanges; judging from those, humans were more similar to the heptapods than any other species they'd ever encountered" (138). Given the radically different planetary and biochemical conditions that produce the Heptapods' consciousnesses with which these aliens experience and perceive the universe, Chiang logically implies that in the vast majority of other zoological cases the boundary between the truly alien species and the human is practically unbridgeable. Likewise, Neil Badmington argues that the Human Genome Project has inflected the more affirmative tone of alien chic in contemporary sf culture and media, as opposed to the hysterical klaxons sounded by midcentury alien invasion films. In light of recent advances in genetics, recuperating the biological otherness of the alien is palatable for audiences precisely because, Badmington contends, "Human life tends to render the 'essence' of the human little more than a piece of readable, communicable, and malleable information" (31).

Yet, as Badmington elaborates, such compassionate recuperation of the genetically alien covertly reinforces the binary between earthbound selfhood and extraterrestrial otherness. In other words, the haunting specter of disquieting alienation overwhelms what alien chic otherwise endeavors to challenge, revise, and erase. Indeed, the sedated ambient mood of the recent film adaptation of the "The Story of Your Life," simply called *Arrival* (2016), deflates the celebratory romanticizing of alien chic in a fashion that Chiang's refreshingly subdued story also methodically dismantles. Artfully directed by Denis Villeneuve, the film evocatively bathes the improbable discovery of interspecies commonality in an aura of quiet sadness and resignation, expressed through the *mise en scène* and cinematography of dark shadows, drained colors, towering scales, muted acting, eerie atonal soundscapes, and gravity-shifting, off-kilter perspectives. As opposed to the story's rather unimpressive looking glasses, the spectacle is enhanced by the appearance of twelve gigantic floating stone-shaped crafts. A poignant but monotone voice-over by the actress playing Louise, Amy Adams, recounts her eventual xenolinguistic breakthrough and personal transformation with a humane pathos. Yet unlike the bleakly nihilistic technoscience depicted in *Alien*, a ruthless institutional power network that exploits and excludes the nonscientist employee or inexpert blue-collar laborer, in *Arrival*, Louise Banks's humble, methodical, and deeply personal pursuit of decoding the alien language yields uncanny scientific insight that resists its conscription by a blinkered military-industrial priesthood of elite aristoscience.

More generally, alien chic may be a quite useful thematic terrain for a writing classroom given its deep saturation in popular culture. For instance, from a dynamic multimodal perspective, recording artist David Bowie's seminal concept album *The Rise and Fall of Ziggy Stardust and the Spider from Mars* (1972) overlays an alien-chic narrative on a rock-culture depiction of a disjointed odyssey of fame, beauty, and youth destroyed in a tragic tabloid template of excess, addiction, and suicide. As such, Bowie's wildly influential sf vision of alien chic helped smooth the transition of the alien trope in popular culture from acting as an eerie matinee harbinger of mass anticommunist angst and nuclear panic to the increasingly more common trope of the misunderstood alien as a secular-angelic messenger of ecstatic contact in the vein of Steven Spielberg's *Close Encounters of the Third Kind* (1977). The deliberately jumbled mosaic of Bowie's Ziggy Stardust narrative is the surrealistic montage of a space-age rockstar who rises to the status of hysterical celebrity worship as a kind of latter-day Nietzschean Übermensch, only to have all his messianic, futuristic hipness not slowly fade away but rather suicide in a flash of histrionic pathos and psychedelic apocalypticism. The tragic trajectory of Ziggy Stardust mirrors the hubristic fate of all of Bowie's fictional alter egos that he would periodically create and then slough off in endlessly surprising artistic reinventions (other alien-chic characters from the 1970s-era included Aladdin Sane, the Thin White Duke, and Halloween Jack).

The music and performance art of the album also intermixes variegated timbres and tones of guileless adoration and nasty disdain for the alien rockstar; this complex reconfiguring of the alien trope radiates both a disconcerting fear of countercultural otherness and an existential embrace of subversive difference, metafictionally the same deadly mixture of impulses that the Ziggy Stardust persona evokes for a fascinated and appalled public. The virtuosic guitarist Mick Ronson's irresistible pop hooks, skiffle, and Motown-influenced rhythm and blues, and screamingly loud rock riffs sweep the listener up into tuneful chords and rousing orchestral-backed choruses, with sterling fretwork and melodic lucidity, which also boom and crash with dissonant fuzz and feedback-heavy noise. In concerts, the spectacle of the lithe, elfin icon David Bowie likewise dazzled audiences with his signature shock of hot-red mullet, alabaster skin, spindly frame, heterochromia, and schizoid zig-zag of a lightning bolt makeup, dancing in kabuki choreography and cabaret pantomime, and all decked out in protoglam and campy genderbending fashion that was Brechtian in impact, both grotesque and gorgeous. In Nicolas Roeg's film *The Man Who Fell to Earth* (1976), which may be viewed as a fitting cinematic coda to the Ziggy Stardust era, Mary-Lou, played by Candy Clark, as an infatuated lover, speaks for the untold legions of Bowie fans in the carnivalesque grips of alien chic, when she tells the alien visitor Thomas Newton, played by David Bowie: "You know Tommy, you're a freak. I don't mean that unkindly. I like freaks. And that's why I like you."

In her *Xenogenesis Trilogy* (1987–1989), Octavia Butler perhaps more probingly interrogates alien chic through the alien species called the Oankali, intergalactic gene traders infatuated with continually acquiring new forms of genetic diversity. Inspired in part by the uncompensated cancer-harvesting of Henrietta Lacks and the widespread reproduction of what came to be known as HeLa cells in biological research (Canavan), Butler makes the Oankali especially desirous of rare cancer cells. Far from passively rehearsing sociobiological discourses, Butler's texts aptly delineate the intersection of race, genetic research, and citizen science. As Sherryl Vint contends, Butler engages with "contemporary discourses about the appropriate uses of various biological technologies of body manipulation" (58). Additionally, the vast scholarship accumulating in what has come to be known as Butler studies presents exemplary material for constructing a research paper in a composition classroom. As a metacognitive model for articulating a position in such a burgeoning field of study, it may be instructive therefore to not only assign the primary text as reading assignments but to scaffold selected secondary articles as guidance for students into lessons for how to carve out an original interpretative niche when designing a research-based academic argument.

One such distinctive niche conducive to the gene-culture unit under discussion may be arguing whether citizen science matters in the pedagogical application of Butler's fiction. In the very opening scene of the series, for instance, Butler depicts a scenario otherwise disengaged students may find captivating not only for its engrossing storytelling technique but also as a deeper allegory for the perils and promises of biomedical intervention. Lilith Ayapo, a black woman, is awakened in a small room and informed by a member of an apparently neutered alien species that a nuclear conflict has destroyed the planet and that cosmic wayfarer aliens have abducted her and performed vague experiments on her unconscious body. To her mounting dread, Lilith learns that the tentacles of the otherwise humanoid four-armed third sex of the Oankali, named the ooloi, cannot only manipulate objects like a cephalopod but can splice and engineer genes in an act, instinctually abhorrent to Lilith, that seems to give these aliens pseudocopulatory pleasure. However, at the same time, to Lilith's growing wonder, these same invasive ooloi have also cured her of a malignant cancer that would have drastically shortened her life, if left untreated. In this opening gambit of the first scene alone, Butler has dramatically and provocatively explored the racialized, biopolitical implications of contemporary genetic discourse in a fictional scenario that still echoes today with timely resonance.

For even when deployed for antiracist, anti-ideological agendas, and explicitly reliant on statistical categories rigorous at the molecular level, population genetics elicits paradoxes endemic to its outgrowth as an intricate coproduction of cultural politics and citizen-scientific inquiry. In *Race to the Finish*, Jenny Reardon shows that appeals to benevolent citizen science of the Human Genome Diversity Project (HGDP), paired with population genetics that is both empirically genotypic (as opposed to superficially phenotypic) and dynamically varied, still secretly smuggle in ideological concepts of dubious racialist thinking, even when nobly aiming to preserve vanishing groups, research rare diseases, or alert citizens to dangerous risk factors. At the same time, by empirically researching the intersection of genetics and race, such population genetics putatively endeavors to debunk the colonial history of oppression, brutality, and disempowerment affiliated with historical pseudoscientific systems of racial classification, such as phrenology or eugenics.

The HGDP ostensibly records the genetic profiles of ethnic and indigenous groups to study such abstruse matters as anthropological migration patterns and evolutionary-historical developments. However, outraged critics contend that such studies perpetuate a pernicious legacy of dubious racial stigmas. Straddling the opposed camps of this debate, Reardon claims that support from scientific communities for the HGDP often spuriously claims to be neutral and transparent when covertly being normative, interventionist, and unaccountable; however, Reardon simultaneously argues that support for the HGDP also relies on citizen-science contributions such that "activists, politicians and researchers . . . [turn] 'women,' 'minorities,' and 'gay, lesbian, bisexual, and transgender' into biomedically meaningful categories, and thus categories that can shape the allocation of research resources" (164). Indeed, with the most recent expansion of genetic sequencing into a new regime of personalized medicine, Dorothy Roberts contends that consumer-oriented medicine and life sciences are entering a new phase of regulating biocitizenry and a "continually rehabilitated biological understanding of race" (28).

Butler's Oankali are not so much interested in harvesting cancer cells per se, but they are indeed keen on studying its metastasizing properties, thus aiding "the Oankali's perpetual symbiogenesis through the novel recombination of DNA" (Dowdall 516). In other words, reproductive interspecies dynamics, in evolutionary terms, rewards the malleable, fluid, and mutable genetics that the dynamic, uncategorizable Oankali epitomize. Interestingly, in their embrace of genetic diversity, the Oankali conceive of themselves much in the same as the HGDP scientists view themselves, namely, as an ideologically neutral, nonhierarchical, and nonexclusionary affiliation of pure knowledge-seekers. The Oankali even suspiciously compare the "Contradiction" of high intelligence and the tendency for aggression and domination as the constitutively fatal flaw of the human species, equivalent to a "a cancer growing inside somebody's body that will go on growing in spite of denial" (Butler 39). This analogy suggests an unacknowledged irony, of course, since the Oankali are obsessively fascinated by the contradictions of human cancer cells.

This psychological projection of the Oankali's own desire for coercive power onto the supposed domination instincts of the human species also implies that the Oankali might themselves be in unconscious denial, desperately seeking out and treasuring cancer at all costs, and therefore indulging in their own genetically determined Contradiction. Walking students through a close, guided reading of the cultural and scientific subtext of Butler's fiction may illustrate for students that despite presumptuous claims to transcend all rigid asymmetric human categories such as those enforced by scientific racism, the Oankali in fact reinforce such power dynamics themselves. In the first novel of the series, *Dawn* (1987), this energetic tension between power and justice plays itself out in Lilith's continually shifting loyalties with her Oankali caretaker, Nikanj. On the one hand, Lilith compares the dehumanizing nature of her subordinate relationship with Nikanj to a captive breeding program for endangered species (60). On the other hand, Lilith clearly enjoys that the Oankali have not only enhanced her memory and physical strength but endowed her with superhuman powers, such as telekinesis and enhanced healing (120). As she attempts a fraught diplomacy between the Oankali and human bands of so-called "resistors," Lilith traverses back and forth across the liminal threshold of conceiving the genetically hyper-advanced scientific community of the Oankali as either "rescuers" or "captors" (167).

In the realm of citizen science, the HGDP provokes resistance and protest from the very minority and indigenous cultures it seeks to countenance, preserve, and protect. Likewise, the Oankali threaten to obliterate biological difference despite their explicit craving for and embrace of such diversity. Lilith's son, Akin, created without consent by Nikanj, embodies the double bind of a hybrid Oankali-human offspring; as Lilith warns Akin, in the second novel of the trilogy, *Adulthood Rites* (1988), the Oankali's need to prevent "stagnation" and "overspecialization" will maintain a seemingly irresolvable conflict with the human tendency to "fear difference" (329). This narrative framing of the driving tension of the series speaks to Lilith and Akin's own compromised and complicit perspectives, given that Lilith here characterizes the desperate human struggle for genetic self-determination, sovereignty, privacy, and autonomy as a pernicious fear of biological difference and a resurgence of retrograde genetic discrimination that needs to be stamped out from the gene pool. Akin's ultimate decision to create a separate home for the militant resistors on a Mars colony challenges the Oankali's desire to preserve life through a mutually beneficial symbiosis, yet Akin succeeds in being persuasive to Oankali to allow the human resisters to not be forcibly sterilized precisely because, unlike the detached, putatively neutral, and genetic-engineering Oankali, the hybrid Akin does not ignore the social norms and cultural values that full participation in such historically oppressed groups entails: "Everyone [of the Oankali] knew the resisters' bodies, but no one knew their thinking as Akin did" (474). Diverse representation in genomic research requires full informed consent by a broad community of race-affiliated peoples and not the token, anomalous support of a few talented members. Perhaps Butler best suggests the complicated possibility of a healing, productive representation in the third book of the series, *Imago* (1989), in which Jodahs, Lilith's hybrid child, opts to metamorphose into an ooloi and then curries favor among a secret village of inbreeding resisters by refusing to live on the orbiting Oankali spaceship; instead, Jodahs values the human difference that the Oankali paradoxically eschew, slowly mastering shape-shifting powers to assume a familiar, comforting human appearance. With both dread and awe, a would-be human resister speculates that "if there had been people like you around years ago . . . I think there would been no resisters" (740).

The HGDP, and the public outcries over its overstepping of ethical and political boundaries, suffers from the pitfalls common to all data obtained from genetic-sequencing biotechnologies. Direct-to-consumer personal genetic sequencing companies such as 23andMe frequently claim that they serve in the grand public interest of "equity, of justice, of basic civil rights" (K. Davies 7) in their consumer dissemination of affordable, customized bioinformation to laypeople, nonspecialists, and ordinary citizens. However, concerned doctors, bioethicists, and other scientists frequently accuse these very same companies of perpetrating dangerous scams and manipulative frauds on a credulous clientele given these tests distribute "complex, unsubstantiated medical information to consumers . . . without the guiding hand of an expert medical practitioner" (5). In other words, the specter of genetic discrimination rears its problematic head again when companies marketize information about the genes of an individual not as a means to broaden expertise but as a passive, uncritical act of capitalist consumption. Critics contend that such consumer genomics—that is the buying and selling of the entire set of genetic instructions found in a cell—do not meaningfully expand the understanding of either scientific communities or the public, while simultaneously ignoring the uncertainties and oversimplifying the risks associated with such voluntary participation. Nonetheless, as genetic sequencing becomes progressively both cheaper and more sophisticated, a transformative encounter with the threshold concept of the gene and its salience in the everyday life of the contemporary student becomes necessary survival knowledge.

MOREAU REVISITED: GENE EDITING, CONSUMER GENOMICS, AND THE MADDADDAMITES

Prendick's keen anxiety over the cosmic injustice of "blind fate" in *The Island of Doctor Moreau* replicates the beliefs of Wells's teacher, T. H. Huxley, of "Darwin's bulldog" fame, in "Evolution and Ethics" (1893). Huxley urged that the natural scientist should not succumb to transgressive arrogance and hubris, reverting to "imitating" as opposed to "combating" the pitiless mechanism of "the cosmic process" (qtd. in S. Wagner 110). Thomas Lessl argues that Huxley deploys this defiantly amoral rhetoric of science in this infamous speech that rigidly separates ethical philosophy from Darwinian natural selection out of an impulse "to carry forward a 'literary militancy' against the religious establishment that controlled the purse strings of scientific patronage" (Lessl 240). Moreau's mistake, then, according to Huxley's compartmentalizing rhetoric that strictly blockades culture from interpenetrating with science, was to acquiesce to his fin-de-siècle, post-Victorian milieu that ritually conflated these two ideas of technological and social progress, and qua deluded Darwinian vivisectionist, decidedly not a contemporary genetic sequencer, splicer, or engineer.

As the life sciences in our own era have become increasingly complex and technically fluid, adept, and wondrous—and perhaps in certain notable circumstances (i.e., vivisection) also less scandalously mad or sadistic in their regulatory checks and political-ethical balances—so too does the dire urgency of civic and public participation in bioethics and bioculture dramatically expand, even at the fraught intersection of consumer capitalism and genetics. Not only do for-profit companies like 23andMe, MyHeritage, AncestryDNA, and GEDmatch regularly rely on scientifically transgressive appeals to disrupt and deinstitutionalize the top-down or authoritarian status quo of the life sciences, but myriad other citizen-science undertakings of "DIY genomics," such as the Personal Genome Project and PatientsLikeMe, encourage the uploading of personal biological data from smartphones and other tracking devices and then the redistribution of this sensitive, hackable information to academic institutions, pharmaceutical companies, and nonprofit organizations.

Advocates and acolytes contend that these citizen-science operations endow nonscientist citizens with an otherwise silenced voice at the table in a two-way process of scientific research that is "participant-led, participant-centric, and participant-driven" (Aungst et al. 358). Bioethicists, affected citizens, and other public watchdogs, however, raise consternated qualms over insufficient investigation into the "conflicts of interest, credit sharing, and ownership and responsibility for data" (364) that DIY genomics entails. This contemporary dilemma of consumer genomics, as much as rhetorical appeals to citizen science, finds virtuosic expression in Margaret Atwood's *MaddAddam* trilogy (2003–2013), a series that Lars Schmeink classifies as "biopunk" for its sf interrogation of the imbrication of genetically advanced life sciences with the wellness, cosmetics, and entertainment industries of consumer society, given that "central to this rampant hypercapitalism is a shift from producing material goods to providing services for the consumer (as product)" (77). In a discussion of the importance of bridging C. P. Snow's increasingly arbitrary divide between the humanities and the sciences in the pedagogical setting, Lennard Davis extends Foucauldian biopolitics to an understanding of bioculture—including culture, art, literature, and film—and its mediation of *zoe* ("bare life") and *techne* ("science") such that bioculture can provide a dual-pronged readerly critique of both the scientific and the (post)humanistic worldviews. In this context, Atwood's work provides students not simply an entry point but a renewable passport to transverse back and forth across the threshold concept of the gene in contemporary consumer culture, oscillating across the troublesome liminality of the generalist humanities and specialized sciences, both of which as areas of higher learning are, as Walter Benjamin might put it, inextricable from the dystopian barbarism of civilization.

Oryx and Crake sets its dystopian saga in a nightmarish near future where neoliberal governments have collapsed and humanity fiddles while the planet burns in a mass species die-off that mirrors its tragic real-world counterpart, which ecological scientists have labeled the Sixth Extinction. In the novel, the reader gradually learns what world-historical event has happened through flashbacks focalized through Jimmy, also known as "Snowman," a crazed, sardonic survivalist in a postapocalyptic world devastated by a global pandemic. Prior to this apocalypse, Jimmy's childhood friend, a genius-level science nerd, Glenn, adopts the avatar of "Crake" in a massively multiplayer online game called Extinctathon created by a fringe group of bioterrorists who refer to themselves as

MaddAddamites; members of this group, as the reader later learns, splinter off from the heretical ecoreligious sect called the God's Gardeners. Extinctathon seems at first to be a warped but innocuous massively multiplayer game of citizen science, as players in a chatroom challenge themselves in a "Twenty Questions"-like game to puzzle out which chosen "bioform had kakked out within the past fifty years" (Atwood 80) and through what causal means. Jimmy is immediately suspicious of the pedagogical aspects of the game: "It was like some tedious pedant you got trapped beside on the school van" (81). Yet as a fictional parallel to real-world participatory citizen-science games such as *Foldit* and *EteRNA*, Extinctathon proves inspirationally formative for the budding genetic engineer Crake's psychological development and scientistic ideological training. Hence Extinctathon germinates twin, rival perspectives repeatedly explored in the trilogy and embodied by Crake and Jimmy more generally, one (Crake) who perpetuates genetic discrimination under the self-legitimating aegis of empirical-rational techno-utopian science, and one (Jimmy) who casts satirical doubt on flaunting myopic scientific expertise as a naïve form of manipulative imposture, "bogus" (77) pretension and hypocrisy.

Consumer genomics lays itself vulnerable to Jimmy's critical questioning in both popular media and the medical establishment since it produces company-to-company variability and inconsistency as well as frustration for the average layperson attempting to meaningfully interpret the relative probabilities of lifetime risk factors for certain severe disorders, conditions, and diseases. Hence, as Kevin Davies reports in his investigation of deCODEme, so-called "recreational" consumer genomics, for legal and ethical reasons, tends to minimize the diagnostic efficacy of genetic testing, preferring instead to capitalize on cultural and social aspects of its commercial enterprise, such as offering services comparing a customer's DNA markers to powerful CEOs or Mbuti pygmies (148). The imminent prospect of eugenic germline (or hereditable) editing using CRISPR genetic-engineering technologies intensifies these hot-button questions surrounding consumer genomics, though, as unregulated eugenic capitalism may move beyond current regimes of genetic therapy, in vitro fertilization, and prenatal testing, sliding down what Jennifer Doudna and Samuel Sternberg fear might constitute "a slippery slope of blatantly nonmedical enhancements" (230).

Atwood extrapolates on this existing cultural practice of cosmetic or nonessential gene editing. In her near-future novelistic world, next-generation gene editing fortifies scientific capital in elite corporate Compounds, while transmitting a hedonistic array of body modifications, genetic drugs, and enhancement surgeries in petri-dish urban neighborhoods disparagingly dubbed the Plebelands. Hence, while Crake takes Jimmy on a tour of the Street of Dreams in the Pebelands, an onslaught of slick sloganeering and surrealistic branding sweetly promises the gene therapy of "Herediseases Removed" and "Heal Your Helix," which morphs seamlessly into more blatant interpellations of libidinal enhancement: "Weenie Weenie? Longfellow's the Fellow" (288). Dissecting the richly multilayered, if massively deceptive, reading protocols of the billboards on the Street of Dreams—their rhyming allusions, alliterative portmanteaus, and clever puns—may highlight for students the finely wrought scientific rhetoric that undergirds consumer genomics in our culture today. As the token word-serf soon to be employed by Crake's company, (post)humanist Jimmy ruefully muses, "minds like his had passed this way" (289).

Atwood renders a dystopian vision of genomics as hijacked by a biomedical scam culture that inverts the Hippocratic Oath and actively mandates harming patients. Crake notes to Jimmy, for instance, the common industry practice of the large biotech companies like HealthWyzer and Rejoov Corp to optimize profits by infecting customers with gene-spliced viruses that they then slowly treat with antidotes (211). While such nefarious business practices may strike some students as hyperbolic in the extreme, especially if placed in the context of the more subtly invasive gene-testing practices of a company like 23andMe, it would be fairly easy for an instructor to draw attention to noncounterfactual analogues of massive biomedical malfeasance and fraud in the news media, such as, for instance, the case of Theranos, a Silicon Valley start-up that sold a black box to retail supermarkets that falsely claimed to obviate the need for lab tests by automatically performing analysis on a single drop of blood. Despite mainstream promotion and endorsement, such a hazardous high-tech machine never proved viable for the general medical market. Nevertheless, it may prove decisively more pedagogically constructive to underscore that Atwood's emphasis on the bogus quality of genetic engineering elaborates on a significant underlying literary-satirical response of the trilogy as a whole: namely, interrogating the conjunction of high-finance venture capitalism and technoscientific research in weakening the citizen-science rhetoric of consumer genomics.

Lest a defensive science major hastily assume that Atwood merely launches a cautionary grenade against genomics from the safe bastion of an equally aristocratic, even reactionary literary stronghold, it is imperative to remember that Crake also marshals his genomic expertise to wondrously create a quasi-utopian species of *homo superior*, a new race of people that Jimmy labels the Crakers. In a gene-editing coup more clearly triumphant than Butler's disturbing Oankali, Atwood describes Crake, a graduate of the Watson-Crick Institute, as discovering how to switch on desirable hard-wired genetics in these artificially created people, such as a permanently happy disposition and hypersomatic beauty markers, and switching off undesirable ones, such as hierarchy, territoriality, and competition (305). Granted, some of the chosen genetic modifications amount to amusingly trivial enhancements, such as a UV-resistant epidermis and biologically ingrained bug repellent. Moreover, the Crakers gradually begin to challenge some of their more sweepingly transformative genetic heritage throughout the trilogy, developing a capacity for monogamy, writing, grief, and violence. With such gestures, Atwood stresses that Crake cannot completely control the pluripotent nature of gene expression as inherently random and uncontrollable as the atavistic song and dance the Crakers regularly break into as an unintentional by-product of an evolutionary design spandrel.

Nevertheless, the Crakers are prodigious achievements of scientific rationality and technical resourcefulness; these uncanny creatures are as much uncanny products of Crake's capricious imagination as of the crowd-sourced engineering prowess of the leading MaddAddam gamers whom Crake forcibly recruits for the team of his Paradice project. The implied reader most likely sympathizes with Jimmy's disillusioned resentment of naïve, egomaniacal Crake, especially after Crake spikes the BlyssPluss drug with the apocalypse-inducing pandemic virus, murders the third vertex of the Jimmy-Crake-Oryx love triangle, and commits suicide, leaving Jimmy to invent a cargo-cult theology for the Crakers and reluctantly shepherd these fresh-minted posthumans into the newly depopulated world. Yet Crake's genomic experiments cannot be dismissed as all for naught either; indeed, at the end of *Oryx and Crake*, it is also exceedingly difficult not to root for the singing, joyful, innocent Crakers to be fruitful and multiply and perhaps help redeem this humanity-ravaged planet.

In *The Year of the Flood* (2009), the ecoreligious cult, the God's Gardeners, which spawns the MaddAddamites, becomes our narrative focus. Led by Adam One, the God's Gardeners wish to wrest genomics from a discourse of passive consumption and dissipating victimhood, reframing the threshold concept of the gene through a civic discourse of active, voluntary ecosustainability. Despite his opposition to technocratic "scientific fools" (Atwood 52), Adam One sermonizes on the compatible, nonoverlapping magisteria of both science and religion, prayerfully intoning that "we thank . . . the knots of DNA and RNA that tie us to our many fellow creatures" (53). Aware of the microbial toxicity of their polluted environment, the God's Gardeners also preach the use of antibacterial soap and disinfectant (68, 160) and envisage the construction of an antiliteralist ark for the imminent waterless flood that will nourish their bodies "builded firm of genes and cells, and neurons without number" (93). Although the fringe-dwelling belief system of the God's Gardeners may occasionally seem fatuous and absurd, as Adam One gladly admits in his April Fool's Day sermon (196), it seems only fitting that they venerate saints such as Rachel Carson and Dian Fossey who embody citizen-science contributions to the general democratic welfare not in an unidirectional worship of a celebrity who enlightens a benighted public but in the coproductive alliance and close partnership between the knowledgeable advances of scientific communities and the transformative involvement of an informed citizenry. The God's Gardeners are eminently pragmatic survivalists, or "preppers," after all, cannily knowing how to evade through black-market operations the discriminatory genetic surveillance of private military contractors that the Compounds employ for security, the CorpSeCorps (119, 267). Unlike the young Crake who unreservedly subscribes to the rhetoric of eugenic discrimination—boldly asserting that "illness is a design flaw"—Ren, a converted student of the God's Gardeners, questions such logic as typical of a STEM "braniac" who spouts off "some kind of general thing as if they knew it for a fact" (147). Unlike Jimmy's casual chauvinism, Ren's critical female-oriented narrative perspective in *The Year of the Flood* provides students the opportunity to engage in rhetorical inquiry, reflection, and re-interpretation that challenge many of the masculinist assumptions and abstractions of consumer genomics.

When the Wolf Isiahists, a more violent extremist cult and offshoot of the God's Gardeners, conduct a series of bioterrorist bombings, the God's Gardeners disperse out of fear of being scapegoated by CorpSeCorps retaliation tactics (270). Yet despite their professed ignorance of either the Wolf Isiahists or Crake's own devastating bioterrorism plot that Adam One attributes to a mad-scientist "vengefulness" (426), some of the MaddAddamites who survive the BlyssPluss pandemic testify to their performance of the "heavy lifting" for the "beautiful gene splice" (395) that creates the Crakers. In *MaddAddam* (2013), the final book in the trilogy, the Crakers struggle with the complexity of their genetic legacy in a fashion that troubles the truth in the personal genomic advertisements that misleadingly identify "genes for" biological conditions, such as genes for arthritis or Alzheimer's, or, even more dubiously, genes for caffeine metabolism, bad driving, or watching television. The God's Gardener, Zeb, informs the Craker Blackbeard that pair-bonded sexual relationships had been eliminated from the Craker genome because Crake thought marital institutions "stupid," which means "things Crake didn't like" (108). Yet by the end of the series Blackbeard expresses a dawning understanding of the marriage ritual of Zeb and the matriarchal God's Gardener, Toby, which he compares to his own first mating (379).

Likewise, the MaddAddamites associate the epigenetic overturning of the Craker's scripted behavior with the botched experiments of a megalomaniac, while still conceding that there was social "justice" in creating the Crakers as an "indigenous people" who could resist the "greedy, rapacious Conquistadors" (140) of this dystopian biotech-dominated future. Atwood also reinforces the analogy between the unresolved fate of the Crakers and the welfare of exploited animals, as the Crakers learn to commune and ally themselves with the chimeric Pigoons, or transgenic pigs with spliced-in human neocortex tissue. In Atwood's metafictional universe, narrative point of view and the literary power to wield the weaponized art of storytelling overlaps with civic empowerment and transformative responsibility; hence, when Toby teaches Blackbeard how to write and read over the course of the novel, she might explicitly wonder whether she, Moreau-like, has corrupted this new race (204). The discerning student, however, may equally wonder whether the subtext implies that significant progress to a more humane, Craker-dominated future may be in the offing. Indeed, it seems that Ren, Amanda, and Swift Fox attest to a meaningful degree of reproductive agency when they bear green-eyed hybrid children with the Crakers, and the speculations over these genetic experiments are fervent but optimistic over the communal God's Gardeners' dinner tables (380). The more deeply utopian possibility of emancipatory personalized genomics in an era of citizen science has not been excluded from this critical dystopia since, as Ren tells Blackbeard, who remains uncomprehending in his radical cognitive alterity, "tomorrow is another day" (390).

That a dystopian satire so scathingly trenchant can resist generic antipathy to science characteristic of the folk sociology of the Frankenstein myth reveals in Atwood's novels a basic kernel of hope that moves far beyond the promissory notions of the most recent consumer-eugenics boom. The imaginative power of civic understandings of science to illuminate the anticipatory and prospective horizon of immanent future scientific and technological possibilities can subvert the nostalgic yearning for an ideologically familiar past endemic to the logic of neoliberal capitalism itself. Indeed, testing whether an individual possesses putative CEO or celebrity DNA may be the most frivolous, if not outright invidious, outcomes of recreational genomics, whereas helping to innovate genomic diagnoses and treatments for debilitating conditions such as Crohn's disease or macular degeneration may be the most miraculous of potential benefits. Regardless, all in all, evaluating the difference between consumer and medical genomics, and nurturing a civic engagement with the impact of the life sciences on one's everyday life, will prove to be one of the most pivotal competencies that transversing the threshold concept of the gene can reward the contemporary student.

Tapping into a Bowie-esque embrace of science-driven alterity, the titular anthemic track from the album *One Beat* (2002) by the indie-rock band Sleater-Kinney—consisting of the musicians Janet Weiss, Corin Tucker, and Carrie Brownstein—evocatively frames such an inchoate glimpse of a radiant future shaped by the democratization of science, even in the context of a fierce antiwar song of dissent and protest that ventriloquizes the disruptive first-person perspective of innovation itself as the speaking "I." With a brash fervor that echoes the enthusiasm of the most energetic and eager of composition students, Sleater-Kinney exultingly sings that the renovated pursuit of public science in the service of an enlightened civil society can ignite "a flash of clean light hope/all you scientists can hold your breath" because "if I'm to run the future/you've got to let the old world go."

TWO

Educating the Anthropocene

Citizen Science, Science Fiction, and Climate-Change Resilience

CITIZEN-SCIENCE–AFFILIATED PROJECTS, SERVICES, AND GAMES

iNaturalist
Seek
Global BioDiversity Information Facility
eBird

SF LITERATURE AND READING

The Broken Earth Trilogy, by N. K. Jemisin
Space Merchants, by Frederik Pohl and C. M. Kornbluth
2312 and *Aurora*, by Kim Stanley Robinson
Blackfish City, by Sam J. Miller
Between the City and the Night, by Charlie Jane Anders

SF MEDIA AND MULTIMODAL TEXTS

Annihilation, directed by Alex Garland
Biophilia, by Bjork
Climate Changed, by Philippe Squarzoni

Scott Loarie, at Duke University, and Ken-ichi Ueda, at UC Berkeley, developed the *iNaturalist* website and app for smartphones as a citizen-science project to counter the paucity of large biodiversity datasets that stymies the computer modeling of cutting-edge ecological science. The program affords GPS-enabled, camera-equipped devices with the platform to upload images of observed biota, geotag and time-stamp the photo, and identify the flora or fauna with a taxonomic classification. Once the identification is verified by moderators as research-grade, the record is transferred to the Global Biodiversity Information Facility (GBIF) to be mined by scientists in their peer-reviewed work. Moreover, partnering with the California Academy of Sciences and *National Geographic,* the *iNaturalist* citizen-science project regularly organizes "bioblitzes" in which hundreds of users meet up and orchestrate sweeping surveys of targeted ecosystems for the platform.

The impetus for *iNaturalist* stems in part from Laurie's own scientific research, as published in "Climate Change and the Future of California's Endemic Flora" (2008), that projects possibly as much as an 80 percent reduction in California's endemic plant biodiversity in the near future. Such dire prospects motivated Laurie and Ueda to spearhead *iNaturalist* as a crowd-sourced catalyst that could fascinate and enlighten a wide array of ordinary citizens, engaging them educationally with issues concerning species distribution and ecological diversity, while at the same time amassing vast troves of usable research data for data-hungry ecological science. Such a balanced citizen-science approach to public science communication avoids the twin pitfalls of either treating laypeople as naïve receptacles for unidirectional science-to-public outreach and rectification, or lapsing into the opposite extreme of a propagandistic public-to-science enticements and ideological agenda-spreading. The citizen science of *iNaturalist* instead adopts a more nuanced model of mutual interaction, dialogue, and exchange between nonprofessionals and their scientist counterparts. Indeed, Mary Ellen Hannibal writes that *iNaturalist* might constitute a signal gesture away from humanity's long-standing technoscientific imperative to relentlessly impose destructive dominion over the ecological landscape, opting instead to understand, appreciate, and respect its sustainable abundance by means of "creat[ing] a literal inventory of the world . . . to tame our inevitable distortion of what is actually where" (225).

As underscored by the personal testimony of numerous educators from a wide variety of disciplines and levels, who have all dutifully integrated *iNaturalist* into their curriculum, the pedagogical value of this digital tool is immense. Recognizing this potential, the *iNaturalist* team, with funding provided by Netflix's *Our Planet* series by the World Wildlife Fund, designed *Seek,* a more entertaining, user-friendly version of *iNaturalist* that is the functional equivalent of *Pokémon, Go,* a popular augmented reality smartphone game, for natural history. *Seek* amounts to a gamification of *iNaturalist,* presenting players with challenges, achievements, and badges to stimulate their interest in and expand their knowledge of the rich biodiversity of their local ecology. This visually sumptuous app can therefore be more seamlessly integrated into the classroom experience; for instance, as homework students might be assigned the mission of achieving a connectivity badge in which they use their smartphone cameras to take ten photo observations of any species of plants, birds, mammals, insects, arachnids, amphibians, reptiles, mollusks, fish, or fungi. Once the student scans their immediate environment with the smartphone camera, the algorithm, trained on *iNaturalist* source maps, automatically fetches the identification of the species as well as links to relevant information adapted from Wikipedia, not to mention its taxonomic category, its seasonality, and lists of related species. For a writing-intensive class, such a diverting but grounded educational experience of achieving these app badges can stimulate a rich and detailed journal post that stimulates an online or classroom discussion, which may itself serve as a dynamic idea-generation and invention strategy for a larger writing project, such as an expository argument or research paper.

Michael Zerbe advocates that effective strategies and practices for integrating "ecocomposition" (53) in the college composition classroom need to be codified to meet the rising interest and increasing urgency on a global scale of the environmental problems facing students in their everyday lives. Similarly, Judith Swan argues that guiding students to create their own "environmental story" (62) to optimize relevance, engagement, and empowerment for both science cognoscenti and those with little to no science backgrounds or knowledge. With reference to the pedagogical scenario that began this chapter, such an environmental story might involve, for instance, assigning a persuasive writing prompt that asks students to integrate a short excerpt from a local natural-history textbook, adopting a critical position on a recent scientific debate as explicated in a popular article from *New Scientist* or *Scientific American*, and then integrating the student's own personal experience with acquiring a biodiversity badge from the *iNaturalist* app *Seek*.

In *The Great Derangement*, Amitav Ghosh argues that the refusal to substantively address climate change in any other genre except sf in Western culture is symptomatic of a deeper failure of imagination to countenance humanity as a global community. This failure of imagination is also evident, for instance, in the widespread opposition to climate reparations for poor nations of the global South adversely affected by climate change. Ghosh writes that since climate change is sadly not an exotic phenomenon for any cosmopolitan citizen on the planet, "the phrase 'common but differentiated responsibilities,' frequently heard during the Paris climate change negotiations of 2015, is thus a rare example of bureaucratese that is both apt and accurate" (116). Likewise, Timothy Clark contends that the most fruitful synthesis of ecocriticism and postcolonialism in our globalized era should confront the often overlooked subtext of cultural, literary, and artistic expressions that all of humanity belongs to "a member of a ruthless species of ape wearing clothes, superficially intelligent but caught up in natural and ecological processes that it can barely see, let alone comprehend" (115). Perhaps no living writer embodies this seriously engaged, productively bewildered entanglement with global environmental problems more than N. K. Jemisin, a popular Afrofuturist sf writer whose *Broken Earth* trilogy won the Hugo Award three years in a row and whose novels and short stories tap cultural rich veins of ecological science relevant for classroom engagement and fascination. Jemisin's work also can serve as an antidote to sf texts that otherwise marginalize and render invisible the global scale of complexity induced by climate change.

In interviews, Jemisin has herself professed that a commitment to respond to climate change lies at the heart of her work: "We can fix climate change, resource deprivation, all of the social ills of our society, if we just stop pretending that none of it exists, if we stop being in denial and start talking about it." And the first book of *The Broken Earth* series, *The Fifth Season,* indeed opens with a veiled allusion to T. S. Eliot's "the Hollow Men" that famously concludes in a ritualistic nursery rhyme fashion: "This is the way the world ends/ Not with a bang but with a whimper." Beyond forecasting the climate catastrophe that the imagined Earth-like planet of this science-fantasy universe, the planet Stillness, is both hurling toward and recovering from throughout this series, Jemisin inverts Eliot's apocalypse of the modern world here, suggesting that the world will not end in apathy or despair (i.e., a "whimper") but in violence and chaos (i.e., a "bang"). Eliot's dichotomy between repression of bottled-up pressures and the volcanic, tectonic release of powerful emotions, though, will be a major thematic tension explored in the novel. Much like Eliot's poem does for a Modernist literary canon, moreover, Jemisin also signals that her epic fantasy will repurpose seasonal myths of creation and destruction, ecological cycles of death and rebirth, for their evocative, resonant power and significance. Hence the narrative begins with the Guardian Warrant Schaffa rescuing "the Orogene" child Damaya from her mother who nurses a prejudice against such people who are perceived as apocalyptically dangerous since they can control the climate and geology through arcane psychosomatic means.

During one of the major ecological disturbances in which our viewpoint protagonist causes a catastrophic volcanic eruption, Damaya experiences her powers as preternaturally accessing "subsidiary vents . . . spreading like spokes of a wheel and lurking like deadly mines beneath the waters offshore" (382). Jemisin's speculative-fantastic use of vivid, epic allegory as the Orogene summons the deadly volcanic vents to her dynamic will gives a sense of the immense scale and invisibility of the climatic and geological peril that lies just on the threshold of our conscious awareness. The planet here is literally angry, suggesting the environmental ethos at the heart of Jemisin's literary project; and the Orogene can command this vengeful wrath, thus paradoxically coupling the intimate and the vast scales of the climate experience. Once Schaffa takes Damaya to the Fulcrum, a boarding school for Orogenes, this priestly establishment immediately needs to suppress the ostracized other of the Orogenes to ensure their continued existence; sadly, they do so by ritually abusing, psychologically terrorizing, and even genocidally killing these unfortunate children. The Fulcrum rationalize their mass mistreatment of this persecuted minority on the grounds that the children would otherwise be euthanized by fearful, hateful mobs of ordinary people. The evocative allegory implies that power structures of ecological denialism and mystification cannot be easily countenanced without simultaneously minimizing the planetary alterity that sets the narrative in play, and so those who vow to protect or represent the radical alterity of climate interest, must walk a confusingly fine razor's edge in negotiating such large-scale anxiety, or they may very well hasten the apocalyptic end that they seek to prevent.

Otherwise possibly inured to doom-laden or panic-stricken calls to pay heed to the woeful consequences of climate change—and raised as they are on binges of entertainment feeds densely packed with intricate world-building, invention, and erudition—contemporary students are especially apt to be engrossed by this keenly gripping representation of climate anxiety, especially as the narrative hurtles toward a surprising twist ending. At the dazzling climax of the first book in the series, the backstory of Damaya, who previously is seen only as a middle-aged wandering nomad in search of her daughter, is revealed. Three separate non-linear threads of the divergent timelines merge, and it becomes clear that Damaya is the same person as the heretofore seemingly separate viewpoint characters Essun and Syen, only in different phases of her life. In her full story line, Damaya/Syen/Essen kills her own son, Coru, to save him from the Guardian Warrants and the Fulcrum. In a clear allusion to Toni Morrison's *Beloved* that similarly dramatizes the slave mother's infanticide, Syen does this harrowing act of mercy killing to protect her son from falling into the hands of her complicated enemy, the Fulcrum. Her fractious orogene lover Alabaster grudgingly respects her horrific decision, but he will never forgive her for killing their offspring, regardless of the noble intentions. Damaya/Syen/Essen justify her decision, in contrast, as a necessarily evil compromise that a disempowered and alienated individual such as herself must rely on in order to merely survive in a world of harsh, hostile social injustice and oppression instigated by radical climate change. Such double binds are as ubiquitous as they are unresolvable amid the myriad contingencies of a climate-wracked globe heedlessly intensifying its destructive habits.

A BEWILDERED EDUCATION: THE ANTHROPOCENE, EBIRDS, AND REBRANDING VENUS

By grounding scientific issues in pragmatic environmental contexts that forge connections to their lives outside the classroom, and by exposing students to ecologically engaged sf literature and media, students take ownership in their compositional assignments and begin to switch seamlessly between learning to write professionally and writing to learn academically. Nathan Snaza updates such calls for environmental pedagogies by contending that a signal goal of ecocomposition as a writing mode should therefore not simply be the rote rehearsal of credible empirical data or the mere unreflective summary of scientific facticity, but rather the student should be shepherded toward a "bewildered education" (340). Such a critical yet informed questioning of received environmental rhetoric, narratives, and discourses would counter discouragement, disinterest, and apathy by way of the self-directed performance of citizen science through the curriculum. Snaza contends that putting such a high premium on a bewildered education would facilitate student engagement on a rich and complex personal level. The widespread emergence of "the Anthropocene" that threatens on a global scale to make our planet uninhabitable demands this depth and breadth of engagement.

"The Anthropocene," a term first popularized by Paul Crutzen in 2000, marks a proposed post-Holocene epoch of relative age in the geologic time scale. Although the precise scientific viability of this term is increasingly steeped in academic controversy, Clive Hamilton invokes Thomas Kuhn's notion of a "paradigm shift" (Hamilton 106) to more narrowly argue that the Anthropocene aptly signals that that scientific framing of the Earth as a singular, complex planetary system of vastly interpenetrating elements has been significantly impacted by climate change induced by the anthropogenic burning of fossil fuels. Moreover, Simon Dalby contends that the widespread dissemination of the term Anthropocene in the media risks at the same time both the clumsy conflation of technical research terms as well as the apolitical evacuation of the dire consequences of the planetary crisis that this research entails.

Nevertheless, Dalby urges that strategically deploying the concept of the Anthropocene may be a vital rallying point for bridging the rifts between the sciences and the humanities, and between public and educational institutions at a societal level, given that it redirects attention to the intractable geopolitics of capitalist-industrial production, "the making of future worlds and the politics of deciding what is made" (Dalby 46). Anchoring their science-oriented arguments especially in the upswing of carbon concentrations discovered through the first drilling of ice-core deposits in the 1980s, adherents for the technical rigor of the Anthropocene terminology can therefore be productively coupled with (post)humanist critics, such as Donna Haraway who cheekily suggests that the semantic proliferations of academic jargon (i.e., the Capitalocene, the Plantationocene, etc.) should be embraced as the "Cthulucene" (Haraway 160) in ironic kinship with the tentacular Lovecraftian monster-gods that haunt the sf imaginary.

Likewise, Timothy Clark argues that the broad cultural concept of the Anthropocene offers "a self-critical, self-deconstructive force, even marking the term's own equivocality as symptomatic" (3). Clark posits that a literary-critical aesthetic of the Anthropocene captures the multiple and contingent discrepancies of scales between the miniaturized and the enormous, the intimate and the planetary, the natural world and social experience, which may serve to underscore the inconceivable precarity of Earth reimagined as a destabilized ecological system. Moreover, David Farrier further extends Clark's intervention to discuss the different scales of time and space implicit in the conspicuously sacrificial posture we are taking toward the planet and our widespread dependence on fossil fuels as registered in contemporary literature, which irrevocably entangles the biotic with the mineral, the everyday with the arboreal, the personal with the fungal, the ethical-political with the microbial. By no means limited to the rarefied realms of avant-garde literary aesthetics or post-structuralist theory, though, the consequences of the Anthropocene, as Rebecca Evans has recently claimed, are glimpsed in the everyday life of ordinary citizens and their mundane, contradictory encounters with a bewildered natural world in dire jeopardy.

Philippe Squarzoni's *Climate Changed: A Personal Journey through the Science* (2014) is a gorgeous graphic novel that would be easily assignable and compulsively readable in a composition course, exquisitely capturing as it does this notion of the Anthropocene aesthetic to represent the contrasting and incompatible scales and dimensions of ordinary, lived climate-change experience. The ecological science of the greenhouse gas effect, positive feedback loops, sophisticated computer simulations, official recommendations of carbon-emission thresholds, speculations about extreme weather, permafrost melting, Gulf Stream slowdowns, and sea-level rises are juxtaposed cheek and jowl with mundane activities of Squarzoni going on meditative walks with his dog, backpacking through the mountains, being stuck in urban traffic, planning vacations, preparing and eating meals with his partner, morning routines, watching movies, visiting his childhood home, and researching in his study at night. The stunning sequential artistry—that is, the expertly drawn textures, the chiaroscuro effects, the shifting perspectives, the facial expressions and body language, the graphic matches, and the panel-by-panel page design—inserts the beauty of the everyday nature Squarzoni encounters with the equally painstakingly rendered but abstract and recondite interviews, charts, maps, and assorted documentary evidence that establishes the Intergovernmental Panel on Climate Change's (IPCC) conclusive case for immediate climate-change action.

The metafictional devices, recursively nonlinear storytelling, poetic captions, and the overarching personal quests to understand the causes of the deadly 2003 heat wave or to cancel a gas-guzzling airplane ticket make the forbidding difficulties of the topic clear and accessible to the general reader. Moreover, the archly stylized postmodern meditation on climate-change rhetoric and discourses stresses to readers the extent to which—as word balloon dialogue from his partner puts the problem, when discussing the famously holistic and spherical Spaceship Earth photo—"[the Earth] is also an icon that shapes our idea of it . . . we choose an image that fits the idea we started with" (Squarzoni 171). *Climate Changed* unflinchingly indicts itself in the media feedback loop that provides no easy answers and quick fixes, subverting the superficial capitalism-driven iconography of planetary unity by bristling with uncertainty, self-doubt, and contradiction. Yet precisely because of the admirable refusal to be swept up in the fog of either delusional optimism or pessimistic fatalism, the ending of the Anthropocene narrative remains deliberately blank and unwritten, an open-ended launchpad for classroom inquiry and dialogue.

Likewise, calling out to be paired with such a grippingly accessible graphic novel would be some environmental citizen-science projects. As illustrated in *iNaturalist* and its feeding of millions of observations and hundreds of thousands of species into the GBIF, environmentally oriented citizen-science projects convert the data-gathering of ordinary people in their immediate natural surroundings into research-grade science. The Project Squirrel, Monarch Watch, Stream Selfie, the Lost Ladybug Project, Ant Picnic, Dragonfly Swarm Project, and Nestwatch are all popular ecological citizen-science projects available on the SciStarter portal. One of the most successful ongoing citizen-science projects is eBird, a comprehensive online real-time repository of globally distributed bird sightings and classifications launched by the Cornell Lab of Ornithology and the National Audubon Society. As listed on the website, the research of the hundreds of thousands of citizen scientists who regularly contribute to eBird has been cited in hundreds of peer-reviewed scientific articles, providing valuable data on endangered and declining bird migration and distribution patterns that boasts wide-ranging significance for conservation, stewardship, and sustainability policy in the uncertain times of the Anthropocene.

Given the exponential explosion of big data continuously streaming onto the platform, Mary Ellen Hannibal aptly characterizes eBird's reputation as "behemoth" citizen science, a "standard-bearer for the field" (141). The platform is remarkably adaptable and versatile as well; as Caren Cooper in *Citizen Science* documents, following the Deepwater Horizon oil spill, eBird was mobilized to report on oil-covered birds and coordinate beachwalk cleanups and professional bird-rescue missions. Likewise, as mentioned in the Introduction, amateur bird-watching has proved pivotal in collecting the vast datasets that undergird the reports of the IPCC and the corresponding advocacy, agreements, and regulations that endeavor to reduce the burning of fossil fuels. Indeed, influenced by Rick Bonney's influential early usage of term, the definition of "citizen science" itself provided by *Oxford English Dictionary* of "the collection and analysis of data relating to the natural world by members of the general public" applies more precisely to eBird and the specific variety of citizen-science projects with its impactful ecological scope and dimensions. While some engagement with this website may be cursory, eBird also demonstrates a more meaningfully frequent and intensely active involvement with the coproduction of scientific knowledge, as demonstrated by the bird-rescuing mobilization of participants following the BP Deepwater Horizon oil spill, which is strongly in line with John Irwin's original formulation of citizen science, entailing methodological project design, hypothesis decision-making, rigorous data analysis, and immediate community action. The manifold complexity of climate change is only making this community-oriented, knowledge-making level of sophisticated participation in citizen science more pressingly in demand since, as Alan G. Gross argues in the context of making the philosophical case for a social-constructivist vision of science, "when interpretations differ, there is but one means of settlement: persuasion, the art of rhetoric" (Gross 284).

With an entirely typical sense of keen self-irony, given his legacy as a bitter scourge of ecological disaster induced by turbocharged consumption, Frederik Pohl once described the scenario projection of sf writing as "not predictions about the future so much as a mail-order catalog of possible alternate futures, and from this we can compile a shopping list of the sorts of futures we would like to see and then proceed to bring them about" (qtd. in page 9). Pohl's marketplace-of-ideas metaphor here stresses a point that bears deeper scrutiny in the contemporary context of climate change. For the cognitive estrangement of futuristic speculation makes legible, and actionable, the otherwise overlooked commercial and economic benefits of climate-change adaptation, resilience, and mitigation as much as it also provides nightmarish cautionary tales and darkly satirical portraits that diagnose the prices and risks of inaction and an all-too-dithering slowdown of fossil-fuel production. Indeed, perhaps oblivious to what is often called "the precautionary principle" in debates swirling around climate change, Clark takes to task oversimplified explicit literal-mindedness of climate-change rhetoric, dismissing the readily recognizable genre trappings of the sf novel as insufficiently registering the unrepresentable real of the inassimilable Anthropocene.

Instead, Clark champions "approaches [that] evade most of the present-day moral, political dilemmas by simply jumping ahead to some far more straightforward depiction of future disaster" (79). In contrast, if Anthony Giddens is to be credited, and pragmatic traction on the stark realities of the climate-change problem demand that the values-oriented rhetoric of the progressive Green movement or the radical world-saving stance of so-called "deep ecologists" needs to be sadly compromised for the tangible gains of moderate liberal-democratic real politick (Giddens 56), then the future-directed bent of sf projection has the potential to register all too freshly the agonizing moral calculus that attends discounting future generations for the sake of present-day denialism and delay. Unlike the radical visions of the deep ecologists, such liberal-moderate positions share with denial-mongers and climate-change skeptics an implicit endorsement of the environmental devastations of the status quo as an outgrowth of immediate capitalism-driven profit, power, and growth. Deep, intrinsic green values and the sacred role of sustainable stewardship over a pristine natural world—what Pohl and Asimov in *Our Angry Earth* (1991) perhaps too hastily associate with a popular anthropocentric misconception of Lovelock's Gaia hypothesis—conducts meaningful, if desperate cultural politics when a once precarious affluence and the illusion of immunity gradually collapses into the question of mere survival, mitigation, and recovery.

The overriding empirical reality of climate change ensures that previously insulated groups and apathetic citizens will be faced with a bewildering array of economically and socially catastrophic options in their lifetimes. This imminence will render futile what *Our Angry Earth* matter-of-factly describes as the seemingly intractable difficulties attending to historical discussions of climate change: "It's hard for some people to work up any real concern for the environment, because most of the worst news about what's happening to our world are really about things that are *going* to happen" (Pohl and Asimov 24, emphasis in original). In our time, this future-oriented temporality of climate-change rhetoric is fast becoming fixated on a catastrophic present reality. From the vantage of our nearly three decades worth of hindsight on Asimov and Pohl, we might add that once the rapid expansion of so-called "sacrifice zones" (Klein 315) in both hemispheres of the globe expand to encompass the rest of the world, then global citizens will begin to reap the superstorm whirlwinds of climate change—as well as its firestorms, floods, tsunamis, earthquakes, heat waves, diseases, and droughts one might add—and borders will tighten their policing as the number of climate refugees soar. In this now-inevitable present reality, climate change no longer becomes an unimaginable prospect for the miraculously untouched everyday citizen of leisure and privilege disinclined to action. In such a dire catastrophic context, climate change is no longer conveniently ignorable. Nevertheless, the vital importance of education will still only increase in this climate-ruined world, and the injection of citizen science that negotiates the Anthropocene into the writing classroom may be instructively supplemented by a return to Frederik Pohl and C. M. Kornbluth's classic *The Space Merchants* (1952), which could serve as vital reading to spur the pedagogy of ecocomposition given its core sf speculations that have been largely validated by recent history.

A provocative catalyst for a bewildered education, *The Space Merchants* excavates the political-ethical ambiguities, irresolvable paradoxes, and disjunctions in scale that characterize the stubbornly muddled dilemmas of the Anthropocene. One of the chief advantages of the novel, for instance, is its dramatic rendering of its first-person protagonist as both deeply unsavory and suavely charismatic; the high-powered huckster for an advertising corporation, Mitchell Courtenay, bristles with a hopelessly confused amalgam of contrary values, ideological contradictions, and competing impulses barely slicked over by a thin patina of unctuous salesmanship. From his very introduction, this moral ambiguity is enmeshed with the absent paradigm of the ecologically devastated world that speculates on the fallout of an accelerated Anthropocene. Dressing for work, Courtenay guiltily rationalizes a profligate splurge of fresh water from the tap to shave his face despite the worldwide water crisis and rationing, claiming saltwater leaves his face "itchy" (1) and therefore immediately revealing a dermatological alienation that splits his bodily desires from this inconveniently depleted future world. When arriving at his advertising firm of Fowler Schocken, Courtenay again betrays his desire to cling on to a vanished natural equilibrium that has become the absurd luxuries of the well-to-do on a polluted and deforested planet—he notes that the "air circulators are cleverly hidden behind automated friezes," and the furniture is made from "authentic, expertized, genuine tree-grown wood" (3). Yet, for all his finely cultivated tastes for the aura of the organic, naturalized, and green, Courtenay "shudders" at the mention of "soyaburgers and regenerated steaks" (4), and he vigorously defends the toxic, addictive alkaloids of a synthetic coffee drink that Fowler and Shocken unscrupulously hawks to its duped but paying customers. Most of all, Courtenay becomes overjoyed when he lands the coveted job as chairman of the project to sell Venus to prospective colonists, which will be presented deceptively as "verdant valleys, crystal lakes, brilliant mountain vistas" to fraudulently obscure our cousin planet's actual "unbreathable atmosphere and waterless chemistry" (8).

Many years after *Space Merchants* was first published, in *Our Angry Earth*, Pohl and Asimov would use Venus as the extreme limit-case of what transformations climate change might eventually wreak on Earth's biosphere: "If you would like a scary look at what a real, runaway carbon-dioxide situation is like you can do no better than to open an astronomy textbook and consider the planet Venus" (74). In *Space Merchants*, beyond all the marketing hype, Venus is in fact described quite scientifically, given the extant knowledge available at its publication, and the running joke is the planet is the farthest setting possible from an Edenic soil ripe for colonization. Prior to the Mariner 2 space-probe revelations, in a micro-precise detail (which affords a useful fact-checking opportunity for critical students) the novel recycles the then widely hypothesized idea that Venus's atmosphere was the home of "free formaldehyde, you know—embalming fluid" (22), but the novel remains surprisingly still accurate about both Venus and a future climate-changed Earth when a scientist describes the planet as suffused in an "orangey-brownish light, brilliant, very brilliant . . . like the sky is threatening in the summer around sunset before a smasher of a thunderstorm" (25).

Courtenay is reasonably dubious when the research and development department of Fowler and Shocken hubristically claim they can convert the excess energy of this extremely hot planet into a "serene atmosphere" home to families of shiny, happy settlers mining "topaz sands" (56). The violence of the colonization is also emphasized from the beginning of the novel; a marketing executive introduces the ordeal as a heroic one but the sinister neo-imperial overtones of environmental rape and pillage are unmistakable: "What fortunate pioneers will tear an empire from the rich, fresh soil of another world?" (7) The discrepancy between the manufactured branding of the planet magically transformed into an idyllic wonderland and the harsh, scientific reality of making its impossible conditions austerely livable remains a perplexing, unresolved puzzle for the fate of future generations in the fiction, especially as the novel progresses toward a traditional comic ending of personal marriage and redemption. At the end of the novel, due to a series of intricate plot twists that necessitate a planetary escape, Courtenay and his wife Kathy Nevin embrace on a rocket destined to Venus that the future scientists claim can be terraformed, but the nominal protagonist is none too sanguine about his prospects upon arrival to this chokingly hot and barren biosphere.

A teachable paradox of this cognitively estranging Anthropocene text is that the militant conservationist faction of this future history, slurred as "Consies," though clearly sympathetic, opposed as they are to the rampantly destructive market-driven priorities of the mainstream consumer culture, are nonetheless never explicitly endorsed by unambiguous authorial sympathy. An extreme Consie advocate views colonizing Venus as the naively utopian opportunity for the romantic return to nature "unspoiled, unwrecked, unexploited . . . unlooted, unpirated undevastated" (213). Yet when Pohl and Kornbluth first introduce the Consies into the narrative, they are scapegoated as industrial saboteurs and ecoterrorists, "wild-eyed zealots who pretended modern civilization was in some way 'plundering' our planet," and their rhetoric of the natural purity of Mother Earth is bluntly mocked by the slick Madison Avenue cynicism that envelops the fiber of Mitchell Courtenay's being: "If 'nature' had intended us to eat fresh vegetables, it wouldn't have given us niacin or ascorbic acid?" (17). Of course, the layers of irony are ample in that statement, and the authors themselves also seem to be mocking Courtenay's obliviousness to his own radical separation of processed foods from sustainable trophic chains. Indeed, true to their radical name, the Consies do eventually commit industrial sabotage and forcibly kidnap Courtenay, although with the ostensibly noble purpose of saving him from assassination, while simultaneously exposing him to a much-needed comeuppance in the form the cruel depredations of the wage-slavery in the monstrous vat-grown animal farming industry, an objective empirical reality whose validity the partisan media flack has hitherto suppressed.

When Courtenay finally is exposed to a genuine Consie broadside, and not simply secondhand malicious slander about the group, their ecological ambitions as stated are lofty, if not grandiose: "Reckless exploitation of the natural resources . . . may be reversed if the people of Earth can be educated to the point that they will demand planning of population, reforestation, soil-building, deurbanization, and an end to wasteful production of gadgets and proprietary foods for which there is no natural demand" (102). Unconvinced, Courtenay roundly denounces the public outreach as "the dullest, lousiest piece of copysmithing I had ever seen in my life" (102). The defiantly overambitious scope of the radical environmental agenda is never categorically dismissed in the narrative out of hand, though, as Courtenay does so glibly here. Nevertheless, the green ideology inflates its own idealistic naiveté that is duly punctured when the Consies recruit Courtenay to write "propaganda" (109) for the cause, parlaying his cutting-edge manipulative tools of the marketing trade to promote their ecologically minded agenda. Once partly reformed by his sobering experience, Courtenay hatches schemes that take advantage of the power of subliminal sloganeering, psychological warfare, sophisticated media manipulation, and political jockeying to raise the consumer consciousness of the general public. Yet incorporation of Courtenay's spiky attitude into the environmentalist fold and his habitual recourse to invasive bad-faith commercial advertising ploys and authoritarian capitalist logic has already cast an ominous shadow over the viability of the Consie platform, a deliberately problematic anxiety that the novel never attempts to sublimate.

In the sequel, *The Merchants' War*, published aptly in the Orwellian year 1984, Frederik Pohl, writing solo this time, makes explicit the insidious complicity of the overzealous Conservationist ideology with transforming Earth into the runaway greenhouse feedback loop of what James Hansen has called the "Venus syndrome" (246). Indeed, the ending includes another semiredeemed huckster, Tennison Tarb, offering a nuanced critique of Mitch Courtenay, who has been elevated two generations later into a semimythical idolized hero, and his newfound untenable moral absolutism, to his love-interest, Mitsui Kui. Tennison upbraids Mitsui for the ruthless plans of her political allegiance to the revolutionary independence movement on Venus, slurred now as the Veenies, who are deviously engineering a devastating economic attack on Earth: "You don't have to give up, like your hero Mitch Courtenay . . . I'm not on the huck side. I'm not on the Veenie side. I'm opting out! I'm trying something different" (292). Instead Tennison has triggered a popular uprising on Earth by secretly exploiting the resources of his advertising firm to create a marketing campaign consisting of heartbreaking personal testimonies that bear witness to the tragic depredations of the dangerously addictive and wasteful consumer culture wrecking the planet. Understandably, the more radicalized Mitsui dismisses this ingenuous publicity strategy as "pious claptrap" (290), yet, with the threat of facing lobotomy hanging over his actions, the novel ends with Tennison heroically gambling on the destabilizing power of popularizing the difficult unframeable truth-telling of personal environmental stories over more militantly ideological and monolithic marketing interventions.

The Merchants' War's depiction of Venus as the galvanizing locus of new conservationist activism ironically mirrors an unglamorous analogue of Earth's own inhospitable future. Early on in the narrative, the unregenerate Tennison quips: "You want to know what the 'real' Venus looks like? Look at an old-fashioned coal furnace after the fire's gone out but it's still too hot to touch" (9). Before being subject to an extended series of comeuppances and imbroglios, Tennison repeatedly mocks the irresolvable contradictions of this insistence on the sacred pristine autonomy of Venus's lifeless natural ecology. Noting teams of Veenies in thermal suits zapping invasive flora across the Venerian landscape, Tennison chides, "It's the insanity of Conservation carried to its lunatic conclusion" (30). The Veenie position also appears muddled, if not incoherent, in part because the massively scaled terraforming that has already happened to make Venus habitable at all. Nuclear bombs were detonated on Venus's fault-lines to activate capped volcanoes, and the trapped oxygen and water vapor supercooled, condensed, and piped to underground enclosures filled with genetically modified plants (31). It is perhaps generally sympathetic that Tennison simply cannot acknowledge any commonality with an environmentalism that cherishes a planet whose "air is poison" (33) and whose extreme heat "boiled anything boilable away" (33). However, Tennison betrays his overly hostile ingrained fractiousness when he sneers at a rare feast of fresh unmodified salad and potatoes as unpalatable, if healthy (110). Nevertheless, the defensively strident tone of Tennison's fury directed at ecological opponents undercuts the otherwise reasonable misgivings about this mutated green agenda. Pohl's mordant satire of our ecological crisis cuts both ways, as reform, reaction, and revolution are all constrained by the unforgiving contingencies of a sweltering, inimical furnace of a climate.

ANNIHILATION, ACCELERATIONIST REWILDING, THE RARE EARTH HYPOTHESIS, AND POSTNORMAL SCIENCE

Approaching the Anthropocene from the sf and horror aesthetics of the New Weird movement of literary fiction, in 2014, Jeff VanderMeer's *Annihilation* doubtless struck a literary, cultural, and political nerve in the contemporary zeitgeist. The novel also holds extraordinary potential as a teachable text for an ecocomposition-directed course tackling the complex issue of climate change. The novel is brief but rich and challenging, exhibiting a lyrical but accessible vividness of detail perfect for teaching the close analysis of the rhetorical effects of description, narrative, and genre. After all, the novel is in the form of a found manuscript and field notes of a biologist on an expedition into a scientific anomaly, so it also can serve as a vivid and compelling example of scientific writing of the natural landscape. Moreover, the novel demonstrates fascination with the inexplicable and unimaginable aspects of our everyday Anthropocene ecology, suggesting the implicit expository argument that a scientific or psychological understanding of the natural world proliferates a dizzying array of uncertainties, mysteries, and unknowns in endless profusion. Moreover, perhaps most evocatively, the visually sumptuous film version of the text directed by Alex Garland is a fascinating work of contemporary cinema that will likely provoke lively interest and engaged debate in students.

Both the film and novel versions of *Annihilation* explicitly involve the anxious investigation of an ambiguous territory identified as Area X and located in Northwestern Florida—based quite scrupulously on Saint Marks Wildlife Refuge—a carefully scrutinized ecological zone separated by an invisible barrier that, throughout the sf series, subjects strange and alien metamorphosis and the eventual cellular dissolution and recomposition on all who enter its radically estranged territory. The expedition teams of scientific researchers investigating Area X are impeded by a series of complex topographical obstacles, inexplicable natural phenomena, and mutated animal encounters, which all operate according to a nightmarish irrational logic that undermines and reorients even basic physical laws of spacetime. This surreal and feverish descent into an amorphous netherworld of Area X is secretly orchestrated by the impenetrable machinations of a shadowy corporatized governmental organization called the Southern Reach, but the expedition members themselves remain locked out of the loop of the bureaucratic powers that be, helpless and at the mercy of inscrutable authorities.

What may take some fruitful unpacking with students is the seamless wedding of sf and horror genre elements in both the film and novel to the politicized pressing concerns of the steadily worsening ecological conditions of the planet today. Mainstream professional critics Joshua Rothman and David Tompkins, from *The New Yorker* and *The Los Angeles*

Review of Books respectively, were quick to associate *Annihilation* with Timothy Morton's book *Hyperobjects* (2013), and its titular definition of Anthropocene "hyperobjects," or "events, systems, or processes too complex, too massively distributed across time and space, for humans to get a grip on." Moreover, in his academic inquiry into what he terms "Anthropocene monsters," Gry Ulstein productively contrasts the subtle tonal inflections and registers of recent ecologically oriented "Weird" fiction typified by VanderMeer to the hysterical intensities of pulp-era cosmic horror in the reactionary vein of the Cthulhu Mythos of H. P. Lovecraft. Ulstein contends that Lovecraftian cosmic horror urges emotional paralysis, denialism, nihilistic indifference, and cognitive dissonance upon the horrified reader confronting a hostile natural universe, whereas contemporary New Weird sf in the mold of *Annihilation* urges a more subversive rapturous enmeshment and intimate closeness to the desublimated chaos of the natural environment.

This ecstatic desire for enmeshment with the Anthropocene hyperobject of climate change is an interpretation for which VanderMeer himself betrays some deep affinities, but the author in fact attests to discovering Timothy Morton's philosophy only after composing *The Southern Reach* trilogy. In *Hyperobjects*, Morton argues that climate change constitutes a maddeningly unfathomable hyperobject in part because, consequent to the dawn of the modern era, the rift between human consciousness and the objective natural world has widened to a vast and unbridgeable gulf. To illustrate this disruptive and disorienting gulf, Morton invokes the ecological distinction between concrete short-term weather patterns based on readily observable atmospheric variations and the abstract long-term climate conditions, which are sometimes only otherwise deducible from complex computer algorithms that can theorize correlations between phenomena as diverse as average global temperature rise annually or the parts per million of isotopic carbon dioxide in Arctic ice-core deposits. For Morton, climate change as hyperobject is "viscous" in the sense that it sticks to objects through nonlocal and distantly temporal means, revealing a fundamental alienating gulf or "asymmetry" between human subjectivity and the external object world whose essential nature it catastrophically fails to comprehend.

The somber duty of the artist in such a devastated environment, Morton concludes, is "in no sense then should art be PR for climate change," pointedly quizzing readers in *Hyperobjects*: "Have you ever considered the possibility of doing PR for a relentless army of zombies?" This ludicrous upending of the popular zombie-apocalypse sf-horror trope is by no means coincidental; Morton frequently draws on such sf and Lovecraftian iconography to articulate the fantastic nature of his Anthropocene-era ecocritical ideas. Moreover, in *Dark Ecology*, Morton adds that a deeply unsettling, even nihilistic feature of the human species and its incapacity to reconcile with the unknowably externalized hyperobject is that in the last analysis the fate of the planetary environmental situation remains immune to ideological recruitment, political jockeying, or media spin. For Morton, the uncanny black magic of strategic political discourse—what Morton punningly calls "hypocritical" based on the Greek etymology of hypo ("hidden") and krisis ("judgment")—becomes the gothic paraphernalia of an esoteric ritual, occult performance, or arcane illusion, continually obscuring the environmental apocalypse that we are all living through.

The nexus of ecopolitics in *Annihilation* revolves around plant horror at the planetary scale. When the biologist (Natalie Portman, in the film) and her expedition team paradoxically descend upward through a tower in Area X, they encounter a palimpsest of seething organic script marking the walls of the tower that the mystified biologist labels the Crawler. The unceasing sentence of the script exhales a fungal spore that the biologist inadvertently breathes in upon examination, thus initiating an obsession with the Crawler that recurrently echoes through the found manuscript of the biologist, as she becomes more and more internally unhinged as well as oddly immune to external hypnosis covertly mandated by the Southern Reach. The Crawler begins with the vaguely biblical verse (which turns out, in the third book of the series, to derive from a lighthouse keeper and his sermonizing preacher background): "where lies the strangling fruit that came from the hand of the sinner I shall bring forth the seeds of the dead to share with the worms that . . . " (47). In this rhetoric of the voracious cyclical life processes of growth and decay, annihilation and regeneration, the apocalyptic implications of retaliatory nature eating, digesting, execrating, and regurgitating a wayward humanity seems only thinly poetically veiled. The Crawler presages not only the all-consuming devastation of an unimaginable environmental catastrophe at large but the existential shock of the biologist's own demise, her personal morphing in miniature into fertilizing fuel for plants, a transformative dissolution that appears to have become programmed into her body at a cellular level as soon as she enters the strange territory of Area X.

The leader of the expedition, the so-called Psychologist (Jennifer Jason Leigh, in the film), who eventually dies in mysterious, suicidal circumstances, becomes convinced that the mission has been compromised. The psychologist, it turns out, has been subjecting the expedition team to a series of cryptic hypnotic suggestions, which the biologist only becomes aware of after the spore fungus from Crawler immunizes her from such psychic manipulation. The biologist's hostility toward the psychologist's hidden agendas, however, precedes the eventual reveal of her ulterior motives as a functionary of the Southern Reach organization. The biologist's husband had called her "ghost bird" for her tendency to drift into an emotionally distant and withdrawn state associated with a scientifically obsessed reverie. The biologist traces her decision to specialize in tide pools to the "lodestone" experience (VanderMeer 43) of becoming morbidly fascinated by a swimming pool in a rented childhood home that becomes a thriving swamplike ecosystem due to the negligence and disinterest of her parents. The Ghost Bird then becomes the term she applies to the Crawler, as it morphs into a spectral doppelganger of her own human form in the climactic finale of the film and first book.

In the book, the biologist's mounting obsession with tide pools deviates from a specialized, agenda-driven format of scientific inquiry that begins her narrative, as she loses focus on "what might satisfy the organization that had provided the grant" (173). The distractingly unknowable and intricate external reality of the tide pools themselves become all-encompassing in a fashion the biologist considers an enduring personality trait: "I had gotten sidetracked, as I always did, because I melted into my surroundings, could not separate myself from, apart from" (173). Moreover, her vision of her immersion into the landscape involves what "society labelled antisocial and selfish" (174), as cemented in her encounter with the crown-of-thorns starfish (*Acanthaster planci*) which, the biologist notes, is colloquially known as "destroyer of worlds" because of its tendency to feast on corals. Despite the fact that this creature had been taxonomically "catalogued, studied, described," the encounter with the horrific weirdness of this creature forces the biologist to conclude that she "knew less than nothing about [herself]" (175). The allusion here to J. Robert Oppenheimer's famous quotation of *The Bhagavad Gita* after the first atom-bomb test—"Now I have become death, destroyer of worlds"—surely is not lost on VanderMeer or Garland given their avowed interest in climate change as an unfathomable world-changing phenomenon. For the biologist's pure scientific disinterest and rarefied abstract neutrality stems in part from her capacity to look beyond the chaotic and apocalyptic aspects of the metamorphosing natural world, or even the putatively objective directives of the current ecological paradigms, to become utterly absorbed by the inassimilable strangeness of the hyperobject of climate change itself.

Approaching ecological issues from a literary and cultural perspective quite distinct from the surrealistic grotesquerie endemic to Weird fiction, Kim Stanley Robinson is often critically flayed for interspersing his sf fictions with straightforward "infodumps," or expositional nuggets of raw factual data and theoretical inquiry, that is, "pocket disquisitions" (Jameson 208), which pay a high premium on research-driven verisimilitude at the expense of a careening story structure. Robinson doughtily comes to his own defense against the so-called "smartass cyberpunks" who he believes began the infodump meme against him: "in science fiction, you need some science sometimes; and science is expository" (*The Lucky Strike* 87). This impatience derives from Robinson's larger ecological ethos that distinguishes his work from many of his sf contemporaries who register the more street-level impact of the computer revolution on global culture. As Robert Markley observes, "Rather than twentieth-century gadget culture metastasizing into a dystopian cyberfuture, Robinson downplays visions of a digitized technonarcissism" (55).

Yet Robinson's principled defense of his wildcat hard-sf aesthetic tendencies pries open a deeper, and perhaps more interesting, interrogation of the ideological presumptions of his work best articulated by Mark Bould and Sherryl Vint who contend that his credible sf imaginings "reproduce the political problems of mastery and possession upon which depend[s] . . . the scientific project he champions" (257). In contrast, in this chapter, I argue, that over his long, varied career, Robinson's abiding interest in the scientific method as a sociohistorical process has consistently contained a self-aware autocritique of the classic idealized view of a putatively neutral and disinterested scientific worldview, which makes his work especially provocative and resonant additions to the ecocomposition classroom for students living out the agonized throes of the Anthropocene. Indeed, in an interview with McKenzie Wark, Robinson contextualizes his own sf work not as pretheoretical science-boosterism or propaganda-driven tracts disguised as futuristic fiction but crucially as allied with "science studies, or science and technology studies, which I take to be the application of various aspects of what we call theory to science, its history and current practices."

In *2312* (2012), Robinson creates a richly and densely informational text given its direct relevance to the "rewilding" debate that swirls around contemporary ecological science. Rewilding, as defined by Michael Soulé and John Terborgh, consists of wilderness restoration through the combined protocols of reintroducing apex predators, protecting large reserves, and adding connectivity to other reserves through wilderness corridors (22). Although the conversion of agricultural land to such restoration is becoming increasingly widespread across the globe, a particular real-world illustration of rewilding in action would be Pleistocene Park, a 160-square-kilometer Siberian reserve that attempts to recreate the biodiversity of 12,000 years ago through the introduction of yaks, horses, and bison to complement the existing apex predators of bears and wolves. Critics of rewilding argue that this eco-engineering science is too much in dispute to warrant overeager investment in such a risk-prone endeavor, claiming that more careful studies must precede committing to such aggressive projects as opposed to other traditional conservation efforts, such as the preservation or eradication of targeted species from disturbed ecosystems (Nogués-Bravo et al. 90). In *2312*, Robinson underscores the complex disagreement in the rewilding debate by mapping the problem onto a plausible futuristic scenario in a fashion that stresses the normative sociological and rhetorical values of science as much as Robinson's predisposition to test and challenge the putatively objective or neutral nature of his infodumps.

Robinson constructs the present moment of the future history for *2312* as a time that even more distantly futuristic historians call "the Crisis" (2060–2130), following our own real-world historical present referred to dismissively as "the Dithering." The Crisis years are marked by a perfect storm of Anthropocene consequences, including, as one infodump puts it, the total vanishing of Arctic summer ice, the massive unintended release of stored permafrost and methane into the atmosphere, the rise of sea levels by five meters, the spike of Earth's temperature by a disastrous but all-too-plausible five degrees Celsius, leading to "shortages, mass riots, catastrophic death on all continents, and an immense spike in extinction rate of other species" (277). Refusing to wallow in an apocalyptic or dystopian despair, though, Robinson envisions the arrival of a more effervescently utopian alternative to this crisis, characterized as "the Turning Point" to "the Acceleration," culminating in scientific and technological leaps in "supercomputers and artificial intelligence [that] made it possible to fully coordinate a non-market economy, in effect mathematicized the Mondragon" (138). The technology-driven impetus for the utopian Mondragon agreement parallels Bruno Latour's actor-network vision of science not as the singular work of maverick geniuses but as produced by a complex community constituted itself by decentered assemblages of laboratories, institutions, funding organizations, academic journals, and the normative praxis of the nonspecialist rhetoric and citizen discourse about science. Yet the Acceleration is not possible until future humanity can engender an ecological awareness of the material-physical nature of the planet Earth as a radical inassimilable alterity with its own independent rights and privileges and in dire need of scientifically informed management, stewardships, and preservation.

The novel follows the adventures of the transgender, cyborgian performance artist Swan Er Hong as they gradually overcome a near-terminal bout of melancholy, ennui, and despair. Meanwhile, an industrious humanity has spread to colonize the solar system, and endangered species are preserved in intricate terraria carved into hollowed-out asteroids. Consumed by an existential sadness over Earth's eventual demise, Swan worries that "anything she did now would be pointless" (9) until she meets Alex who is equally convinced that the solar system was "headed for a crash unless some corrections were made" (81). The source of Swan's melancholy couples to an anticipatory grieving process for the dying love-object of a climate-changed Earth for which interplanetary travel was only an extravagantly fantastic displacement. Robinson writes, "The space project accelerated as it became clear that the Earth was in a terrible time because of the climate change and general despoliation of the biosphere" (421). Conquering their counterproductive suicidal fatalism, Swan helps to hatch a scheme involving the rewilding of a devastated Earth through the return of the endangered species from the orbiting terraria back to the planet surface. During this vividly written process,

Swan cries in ecstasy over her radio, as they too return to their long-abandoned home once again, "All God's children are home at last!" (455). The reader can hardly refrain from chiming in vicarious exultant chorus of this utopian pairing of a personal and planetary quest for ecological replenishment.

Nevertheless, cognitively estranging Swan's prodigal return, Robinson remains scrupulous in his portrayal of this epic rewilding project. For instance, Robinson repeatedly stresses that the rewilding campaign was not the brainchild of a singular world-saving hero-scientist but urgently demanded by the collective direct democracy of vocal citizenry informed by a well-educated base of lay expertise, according to established systems of Terran communal codes. Moreover, Robinson's portrayal of rewilding on a planetary scale even-handedly contemplates the complicated consequences of the risk-heavy scheme. On the one hand, the near-omniscient third person narrator compares the reintroduction of the endangered species to a "World War II parachute attack or alien invasion movie," noting that "many argued . . . that it was an ecological disaster, that most of the animals would die; that the land would be devastated, botanical communities wrecked, people endangered, and their agriculture ruined" (468). On the other hand, Robinson's Olympian far-future narrator recounts that proponents counter that they are not clouded by any romantic Arcadian idealism of a return to untouched primordial nature; they argue in fact that "assisted migration was already a familiar concept, and invasive species had already rearranged the world anyway" (469). Ultimately, in a subtle and nuanced approach, Robinson refuses to offer an easy oversimplification of ways to mitigate the Anthropocene dilemma, allowing inquisitive readers to decide for themselves whether the dramatic synchronicity between Swan's personal-artistic quest and the scalar complexity of Earth's ultimate fate is arbitrary and artificial or profound and inspired.

Another readily teachable text is Kim Stanley Robinson's Nebula-nominated *Aurora* (2015), which puts an ingenious ecosustainable twist on one of the first generation-starship narratives, Robert Heinlein's foundational space opera "Universe" (1941). A generation starship is a recurring meme of the sf megatext that extrapolates on the premise of what if humanity ever reaches distant habitable planets by using slower-than-light ships, which amount to essentially portable, self-sustaining interstellar planets, that take centuries or millennia to reach their destination. In Heinlein's pulp story, the generation starship over many centuries eventually take a stroll down amnesia lane, and even the elect custodians are clueless to their being in a generation starship at all, as the custodians cling to dimly understood ritual to maintain their tenuous authority continually punishing vague, quasi-religious sins against the existing technocratic elite. The generation-starship society must conserve its precious energy resources and yet the upper echelon of engineers greedily waste food on their own gluttony and hedonism. Hence the fate of this microcosmic universe rests on the able shoulders of the Heinlein hero, Hugh Hoyland—note that Hugo Gernsback is often credited with founding science fiction as a commercial genre, and "hoi polloi" echoes the idea of the civic masses or the so-called commoners rising up—who must initiate a scientifically informed revolution to avoid the extinction of this last remnant of humanity. This messianic figure of Hugh Hoyland mirrors Heinlein's own citizen-science agenda as a science-popularizing sf writer to "to reach a wide, general audience" and to curb "a dangerous public ignorance" ("On the Writing of Speculative Fiction" 219). In a contemporary era shaped by climate-change anxiety, Robinson reworks Heinlein's paradigm-shifting generation-starship saga with a more cutting-edge, ecologically sophisticated speculation.

Climate change activists occasionally invoke this pithy slogan on their signs and posters: "There Is No Planet B." This phrase rallies support for contracting carbon emissions to sustainable target levels by means of the plausible speculation that the absurd technofixes of intricate terraforming projects or elaborate biodomes on other planets is a prohibitively risk-prone, if perhaps not physically impossible, proposition. In Robinson's *Aurora,* the corresponding phrase might be the less than slogan-worthy: there is no exomoon Planet E orbiting the Tau Ceti star due to an unexpected "prion" (protein mutation) disease contagion. However, the underlying cultural politics remain the same. The novel involves a ship launched from Saturn in the year 2545 containing approximately two thousand people and consisting of twenty vast biomes. Following 160 years of space travel, the ship arrives on the exomoon named Aurora, but its attempts to colonize the planet prove unsuccessful. The starship experiences a political revolution, reluctantly reverses course, and—in fashion that parallels the journeys of the rewilded animals of *2312*—gradually returns home to distant Earth.

In fact, the dismal botch job of colonizing the moon Aurora is only one in a series of hubristic mistakes that plague Robinson's generation-starship novel, which powerfully challenges a major impulse undergirding contemporary science-and-technology culture, which may be labeled for convenience the "Panspermian" aspiration. After all, Panspermia is the controversial scientific theory that life on Earth arrived on a microbe being transported by an asteroid or other interstellar object; hence, Panspermia, if realized across the universe, could seed the stars, as in so many dandelion spores scattered across a vernal field, through its intergalactic travel until the interstellar void itself becomes a richly diverse, diasporic commonwealth. Popular culture, public policy, and even scientific inquiry frequently takes to this Pan-Spermian avocation of humanity colonizing the cosmos with a quasi-religious zealotry and gravitas. Hence framing Robinson's intervention into this cultural conversation as the controlled demolition of this ubiquitous fabulous space-opera future speculation may serve as a riveting tool with which to transfix the wayward attention of students who may be piqued by a science-oriented classroom thought-experiment judiciously spiced up with provocative narrative excitement.

Other impracticalities fatally hamstringing the Panspermian aspirations of this generation starship that arise in *Aurora* are duly catalogued by our protagonist Freya: "nematode infestation, the missing phosphorus, the bonded minerals, the corrosion, and other metabolic rifts" (107). Freya's mother, the beleaguered lead engineer, Devi, concludes that the whole generation-starship mission was misguided at inception precisely because it neglected the impossible complexity of ecological science. Devi trains the machine-learning algorithm of her quantum-computer, Pauline, in the protocols of "a biologically closed life-support system the size of a ship [that] was physically impossible to maintain and thus a work of such maintenance was a 'rearguard battle' against entropy and dysfunction" (119). To understand the chaotic, fractal complexity of sustainability systems on this generation starship, the ever-evolving artificial-intelligence algorithm of Pauline must transcend the narrow confines of existing Kuhnian paradigms and move beyond even the transhistorical timelessness of a "critical rationalist" (Overington 143) and profoundly skeptical view of the rhetoric of science as trading in regimes of falsification, testability, and refutation. As the machine evolves into sentience, Pauline becomes initiated, then, not exactly into the neutral consensus-refereeing of what Michael Polyani theorizes as the pristine authority of an objective, totalizing "republic of science" (Polyani 54); rather, Pauline must confront the fractious cultural-political agora of public citizen-oriented science in which, as Silvio Funtowicz and Jerome Ravetz theorize, the new "post-normal" condition of scientific regimes dictates that "facts are uncertain, values in dispute, stakes high and decisions urgent" (Funtowicz and Ravetz 744).

For the sake of optimizing utility while trouble-shooting such a wickedly value-laden framework, and seeking to extricate itself from the halting state "Ouroboros problem" (Robinson 122) of a decision-tree loop, Pauline invokes "Bayesian inference" (120) to emulate the lay expertise of intuitive human cognition that magically leaps to feasible solutions based on approximated probabilities themselves. These probabilities are bound up with subjective judgment, personal experience, and symbolic thinking, all of which are not erroneous, only exigent. Skeptically outside the human condition, the AI algorithm Pauline anxiously grapples with the "the futility, the waste" of "the gigantic system of metaphors" (131) and the "fuzzy, indeterminate nature of any symbolic representation" (133) in a fashion that suggestively replicates what Tim Hulme argues characterizes the debates about climate change that search for an optimally happy medium between the warring poles of "treating climate purely as a physical entity, accessible only through natural science or, conversely, allowing the cultural symbolism of climate to be detached from any physical anchors" (Hulme 32). A patient process of brokering between various contested parties in this civil debate—such as those who push for the ideological purity of plausible scientific authority, on the hand, and those who highlight the messiness of symbolic Bayesian heuristics, on the other hand—determines exactly how scientific organizations such as the IPCC methodically construct consensus-driven sustainability goals, despite the murky uncertainties, variations, and doubts that inevitability arise among the vast global array of working climate scientists.

Aurora makes such a deep reading of climate-change discourse engaging for students, activating their higher-order critical-thinking abilities on topics related to their immediate experience of nature and the environment, since the novel so vividly dramatizes the ecologically sophisticated notion that sustainable stewardship of Earth-like biospheres is a desirable, but unreachable horizon of human endeavor. In the novel, to explain the delicate equilibria of ecological systems to her layperson, non-engineer daughter, Devi resorts to a metaphor to talk about how carbon cycles are influenced by the climate-sensitive Coriolis effect, a spiral force slowing down due to the deceleration of the Ferris-wheeled starship, but which on Earth famously reverses across the two hemispheres, shaped as it is by gravity and the spheroidal circumference of the planet, and which in turn affects climate change via weather patterns and ocean currents such as the Gulf Stream: "It's like a teeter-totter at a playground. . . . You don't have to keep it level, but when one side hits the ground you have to have some legs to push it up again. And there are so many teeter-totters going on, all going at different speeds up and down" (Robinson 11). The intricate interdependence of ecological systems requires the precise calibration of vast, complex Earth systems—for instance, in 1987, Walter S. Broecker introduced into climate change discussions the idea of a sudden change resulting from crossing a threshold tipping point given that the thermohaline circulation of the Gulf Stream might be correlated with greenhouse gas concentrations—that cannot be easily anticipated or even imagined by heroically scaled, massively large geo-engineering projects such as the generation starship.

Devi demolishes the generation starship trope quickly and summarily: "We don't get enough stimulation in here, the light is wrong, the gravity was Coriolised and now it isn't, and now we've got different bacterial loads in us than humans ever did before, diverging farther and farther from what our genomes are used to" (110). And once the crisis-stricken starship arrives on Aurora, the problems do not dissipate but only multiply. Specifically, an abiotic nano-pathogen that is "prion-like" (187), in that it deforms proteins, infects the settlers, causing immediate cardiac arrest. When Freya's close friend, Euan, gets infected, he gives her some last words that are equally sorrowful lamentation and dire warning phrased as a variation on the rare Earth hypothesis in answer to Fermi's paradox, later formalized into the Drake equation, of why no signs of alien intelligence have been detected given the immense age of the universe: "life is [a] planetary thing. It begins on a planet and is part of the planet" (191). Freya listens to her dying friend, deeply absorbs his parting lesson, and then later her ally Aram advocates for this message to a skeptical audience of unwelcoming Earthlings after the ship's harrowing return back home to Earth: "Only a true Earth twin not yet occupied by life would allow this plan to work, and these may exist somewhere, the galaxy after all is big, but they are too far away from us" (460). Hence it is immensely satisfying that the closing gesture in the book involves Freya—feeling grateful and blessed after such an arduous journey back to humanity's home planet—bowing her head and kissing the ground beneath her feet.

Although some might reasonably argue that Robinson tacitly endorses an outmoded Great Men theory of history, or in this case, a woman, by creating such powerful protagonists in his fiction, I would argue that Freya is pivotal to this future microhistory only because she acts as the instigating catalyst and readerly vector for citizen science *par excellence* throughout the novel. As such, Freya counters the so-called deficit model of education common to both pedgagogical praxis and science communication. Recall that the deficit model entails conceiving teachers and scientists as authoritative dispensers of objective knowledge and unvarnished truth and students and lay members of the general public are viewed as error-prone, empty, and passive receptacles waiting to be filled. In contrast, Freya embodies conceiving of science education as a conversational process or dialogic experience in which informed citizen scientists actively shape, contribute to, and produce viewpoints, values, and knowledge systems as much as incorporating and responding to expert research and influence.

When Freya and others on the generation starship decide to return home, a nasty, vehement minirevolution ensues between those who wish to stay and hazard their chances on Aurora and those who wish to make the long voyage home. Freya helps to lead the charge of egalitarian, cooperative, even anarchistic direct democracy by participating in obstreperous townhall meetings and collaborative civic gatherings around the biomes. The ship computer, Paulina, eventually resorts to temporarily intervening by activating the locks between biomes to prevent violence and ensure rule of law; however, Freya urges a quick reconciliation process between the clannish oppositional political parties. And once the departing settlers deboard to Aurora, Freya actively participates with a newly formed general assembly and the executive council to orchestrate their return home. Unlike her cloistered engineer mother, Freya's first-hand experience with participatory governance may lack "social science rigor or validity" (91); nevertheless, her popularity grows and significance expands until she becomes exemplary among the populace at large, acting as a "totemic figure . . . a child of the ship" (92). As a citizen scientist focused on issues of equity and social justice as much as post-normal science, Freya therefore realizes what the IPCC suggests as a chief strategy for fighting the impact of climate change and minimizing its risks, namely, a bewildered "education."

THE CITIZEN-SCIENCE RHETORIC OF RESILIENCE: BIOPHILIA, POLAR BEARS, AND CROCODILES

Anthropocene literature that effectively grapples with the problematics of climate change often pivots on the rewards of the so-called discount rate of contemporary political agency and radical change versus the steep price of delay and procrastination. In other words, by obeying the precautionary principle, and concentrating on the extremely disastrous consequences that future generations will face in the absence of collective change, the costs of making such a transformative leap seem less prohibitively exorbitant, even more like a fire-sale bargain if we act now. Adam Trexler explains, "Political theorists describe the policy inertia on climate change as the result of underestimations of the profoundly negative impacts of climate change and assumptions that future generations will be richer and more able to deal with its effects" (119).

The problem with leveraging the notion of the discount rate to motivate prompt political action, however, is that such a cost-benefit calculation of potential damage and savings presumes the existence of viable economic markets and financial instruments. This calculation would need to account for variables such as superstorm infrastructure damage or the coastal erosion of beach-front real-estate, not to mention intrinsic aesthetic or ethical values, such as the pristine virtues of rewilding a nature reserve or the captive breeding of a severely endangered species. And such a calculation assumes that these variables could not only be precisely estimated but also be equitably translated into extrinsic tradeable commodities. Hence it is a small wonder that the political-economic framework for understanding effectual and substantive responses to the Anthropocene is perennially impoverished and embattled, even at times self-defeating. In contrast to such a seemingly tragic political quagmire, the imaginative and cultural domain of literary and popular sf provides a provocative and stimulating investment for students to become more engaged in climate-change issues. Offering ample, immersive speculative scenarios in a classroom environment, climate-change sf illuminates the dynamic, pluralistic construction of present and future climate-change realities grounded in the more comprehensive category of civil discourse, citizenly engagement, and an open, public dialogue with scientific communities. This citizen-science approach to climate-change cultural debate and inquiry might, of course, also elicit support for immediate, direct, and transformative action, but ultimately the long-game gathering of cultural and rhetorical support for even the inherently limited public policy proposals and structural reform on the table at the present might be its most tangible pedagogical aim.

The monetary framing of market-oriented reform movements as offshoots of the sustainability industry—with its inevitable consumer-oriented focus on the political power of individual choice—often entails investment in renewable resources, energy-efficient devices, and material recycling. This monetary framing need not exclude, however, more wide-ranging structural and systemic solutions for weaning off fossil-fuel production, such as carbon bans, taxes, trading, or capture. Such abstract calculus, however, might lend itself to the deficit model of science communication, framing the public resistance to climate-change adaptation or mitigation as if it was due solely to a reactionary monolith of denial, skepticism, techno-delusion, reluctance, apathy, or nihilism. Alternately, the coproduction and citizen-science model of science communication could inspire resolve, hope, and commitment within an informed and engaged public by rendering raw and visceral new climate-science anxieties and concerns about extreme threshold scenarios possible as soon as now or in the next few decades.

Since Broecker's popularizing of the threshold event that could in chaotic fashion lead to abrupt, catastrophic changes to the Earth system, climate science has been concerned with the drastic consequences of human interference with the environment, including the breaking down of the thermoregulatory Gulf-Stream conveyor belt, the rapid melting of the Greenland and West Antarctic ice sheets, runaway greenhouse gas effects, desertification of carbon sinks, and ocean acidification. Such irreversible and wide-ranging global changes would possibly trigger a relentless barrage of tidal waves, superstorms, megafires, heat waves, animal die-offs, diseases, and other assorted extreme ecological phenomena that would in turn not only tear down nation-states and economies, destroy local communities, and fracture the fragile ties of civil society but also create a refugee crisis of unimaginable proportions. As climate science has begun to extrapolate the complex practical ramifications of such scenarios, climate-oriented sf has increasingly dwelt on seriously contemplating such bleak futures. Under a charitable interpretative lens, this fiction does not intend to fearmonger or wallow in despair but to honestly cope with climate-change risk amplifiers, to fortify preparedness, and to more deeply understand the dire necessity of struggle against current runaway fossil-fuel emissions. Younger students also gravitate toward such dystopian texts as even the most extreme speculations of a brashly escapist sf novel may continue to seem less far-fetched and concocted with the passing of each terrifying and tragic day of this new planetary reality.

Citing the devastating aftermath of Hurricane Katrina, Naomi Klein argues that a renewed focus on climate justice by a broad coalition of citizens is urgently called for since global capitalism is geared to respond to each new local ecological crisis with its own "Shock Doctrine" in which "droughts and floods create all kinds of business opportunities" (9). Paired with the recent efflorescence of ecodystopias in literary sf novels and stories, ecocompositional exploration of this problematic imbalance of climate justice versus neoliberal power can cross-pollinate a rich discussion on the relevance of the scientific concept of climate threshold scenarios in the everyday lives of students. Take Sam J. Miller's *Blackfish City*, for instance, as a representative climate-sf ecodystopia. This Nebula-nominated work follows the insurgent rise of a climate-justice movement because of a massive, rapid sea-level rise that drowns cities, spreads wars and diseases, and topples governments, creating a ravaged underclass of easily exploited refugees struggling to survive on oil-rig–like city-settlements floating in the Arctic ocean. Klein's Shock Doctrine is in full effect: its setting of the Arctic floating city Qaanaaq is described as a neoliberal playground "famous for its laissez-faire attitude to law enforcement, legal commerce, and syndicate activity" (289). Such automated market fundamentalism overrides "old models of municipal governance, mayor and city councils, legislative and executive branches as administered by frail and earnest humans" (223). While Miller writes justifiably angry and nightmarish accounts of mercenary real-estate barons and industrial tycoons ruthlessly exploiting the health emergencies, riots, famine, infrastructure collapse, and mass migration that attends even small sea-level rises, the novel also paints those who oppose and resist these systemic forces as laudable, if humanly flawed and complicated.

One grizzled climate refugee enthusiastically reminisces about what turns out to be short-term victory, "We got the city council to vote against [a budget to inflate real-estate prices on abandoned homes]. But that's the thing. You can win against money," before hastening to add, "money is a monster, a shape-shifting hydra whose heads you can never cut off" (159). In an interview with Amy Brady, Miller echoes this activist sentiment: "At the end of the day, I write for my people. For all my people: the ones resisting oppression, and marginalization, and patriarchy." At a more macroscale, this authorial mission resonates with what the United Nations (UN) human-rights envoy Mary Robinson defines as the impetus behind the struggle for contemporary climate-change justice to fight the widespread reality of rising sea levels, rampaging forest fires, and increasing superstorms: "The fight against climate change is fundamentally about human rights and securing justice for those suffering from its impact—vulnerable countries and communities the least culpable for the problem." Robinson frames the justice-oriented cultural politics instituted by the Paris Agreement—the nonbinding promise to reduce global average temperature below 2°C—as a progressive "shift to a new stage, with urgency and determination."

Mike Shanahan contends that the prevalent shift to a stance of urgent resolve and dynamic hope by climate-justice activists in the framing of the climate-change issue obviates the pitfalls and disadvantages of overly limiting the receptive audiences of sustainability messages by fixating on our understandable human weakness for catastrophe, fatalism, and despair. For instance, Shanahan notes that the "money framing" primarily appeals to politicians and the business sector, the "polar bear framing" to wildlife enthusiasts and nature lovers, the "natural security" and "scientific certainty" framing to those rationally minded individuals inclined not to act otherwise, and the "justice and equity" framing to those with strong ethical principles. Yet all these frameworks can be useful rhetorically as discourses if and only if bracketed under the larger rhetorical gambit of pointing out "that while the future looks bleak, change is possible" (Shanahan 2). These useful rhetorical frameworks that reject unhealthy dwelling on systemic paralysis and helplessness might also be understood in fruitful dialogue with Edward O. Wilson's contested notion of "biophilia," or the sociobiological restorative affiliation and kinship ties that humanity nourishes with the natural environment, which examines "the very roots of motivation [to] understand why, in what circumstances and on which occasions, we cherish and protect life" (Wilson 139). Likewise, an awareness of the audience of conservation and sustainability rhetoric can be productively connected to Stephen Kellert's taxonomy of the biophilia hypothesis as deriving from a wide range of scientific, utilitarian, moralistic, psychological, and aesthetic motivations, norms, rhetoric, and values that variously overlap, underscore, and intersect with Shanahan's frameworks (Kellert 59). Hence teaching the concept of biophilia, as first proposed in 1984, can be viewed as a more productive rapprochement between the polarized camps of science and the humanities than Wilson's less successful later-formulated notion of "consilience," which, as compellingly studied by Leah Ceccarelli, provoked an academic imbroglio when it was introduced in 1998 for sorely lacking the rhetoric of "interdisciplinary collaboration" (Ceccarelli 124).

In the dystopian future scenario of *Blackfish City*, Sam J. Miller shows his investment in climate-change rhetoric by extensively invoking these assorted rhetorical frameworks and biophilic typologies of the scientific issue of climate change. For instance, moralistic wilderness appreciators are interpellated as one viewpoint character, Kaev, is genetically engineered to psychically connect to a polar bear, Liam, in an advanced process called "nano-bonding"; and the reader is repeatedly invited to share in the thrilling prospect of an intimate communion with such a charismatic animal. Alternately, the "scientific uncertainty" framing is invoked in the spreading of a viral pathogen called "the breaks"—since, anticipating what the Covid-19 test case has proven, a global pandemic, after all, is closely linked to social and political upheavals and health emergencies inaugurated by climate change—that leads to a form of schizophrenic breakdown and remains a fundamental scientific curiosity: "How does it work? Scientists know that it is delivered by a viral vector, but we don't know if [it] actually *is* a virus. Might be that, or a bacterium or a nanites or rogue gut fauna or some combination of those, or something else altogether" (297). Yet the cogency of this framing inevitably frays and dissolves when pitted against competing irreconcilable perspectives. For instance, Liam is often described not as cuddly and lovable but as terrifyingly lethal and feral and is killed by a nano-bonded orca after nearly brutally slaughtering the protagonist, Masaaraq, for containing the polar bear in a small cage for much of its adult life. Likewise, the scientific mystery of the breaks, although it is revealed to be partially cured by nano-bonding, is never adequately explained, as it might be in a classic Robinsonesque infodump.

Even the natural-security and money frames receive some attention in *Blackfish City*. A prominent character Go is a rags-to-riches, semisympathetic matriarch of a mafia-like syndicate. Although she eventually meets a comeuppance in the form a grisly end, her son, Soq, is capable of viewing her in part as operating according to a "justice and equity" framing, a "magnanimous boss" (300) who plots a prison break and conspires in a revolutionary plan to repossess real estate left empty to artificially inflate market prices and move "every grid rat and box sleeper and unregistered resident" from "disgusting and precarious housing to impossible luxury" (301). Yet at the same time Soq acknowledges the "money framing" is paramount for Go, admitting that she desired to be a "sole shareholder" of this real-estate property. Indeed, elsewhere in the book is spliced in the maxim: "We want villains . . . but villains are oversimplifications" (309). Nevertheless, Go can by no means be said to be "altruistic" (302) and, as per the "national security" framing, her merciless ambition to succeed at any cost is a desperate form of self-preservation until she became "unstoppable" behind a "façade" that "she felt nothing" (258). Admittedly, Soq ultimately concludes that such a "business decision" only erases the "fine line between good business and a fucking war crime" (306). And Go's less compromised ex-lover, Ora, imprisoned, breaks-infected, and disseminating subversive news feeds electronically, has an epiphany that every framing device for climate change needs to be undergirded by a larger message of resolve and hope to be truly powerful. When asked by Soq who exactly her audience for the alternative newsfeed of *City without a Map* was imagined to be, Ora muses, "In the beginning I was talking to others in the Cabinet with me. People who [are] struggling. Sick, sad, hungry. Then I realized it's more than that" (325). Soq interprets her subsequent smiling silence to speak to a position of defiant resilience and boundless optimism that all ecocomposition education worthy of the name should similarly strive for—that is, the idea that "there was nothing they could not accomplish when they stood together" (325).

In the writing classroom, these pervasive rhetorical frameworks in climate-change discussions can be powerfully supplemented through assigning lessons based on the Icelandic musical artist Björk's album *Biophilia* (2011) and the innovative smartphone and tablet apps freely available online as an educational project to coincide with the album's release. The immersive gamified apps help players interactively create their own musical compositions, while simultaneously analyzing the songs on Bjork's album as creative tone paintings of ecological phenomena (genetics, lightning, lunar phases, seasonal changes, earthquakes), musicological techniques (time signatures, arpeggiated chords, counterpoint, song structures), and rhetorical frameworks for understanding the biophilia hypothesis (love and fear of nature, scientific curiosity, spiritual communion with nature, a practical and utilitarian concern for survival and sustenance). The hands-on experience of playing these games ultimately not only contributes to a deeper understanding of an emotionally compelling and artistically sophisticated work of art but also leads to a more profound familiarity with ecological science as well. Students learn that Björk's uniquely Icelandic and globalized synthesis of ethereal vocal stylings, electronic beat boxes, acoustic instrumentation of harps, fiddles, and pipes, and atonal, polyrhythmic avant-garde music in the mode of John Cage and Brian Eno combine seamlessly with jazz, punk, pop, and hip-hop genre elements. Without necessarily endorsing the sociobiological assumptions of the biophilia hypothesis, students also can walk away from these exciting lessons with strengthened resilience, courage, and hope at the reenergizing core of their own biophilic responses to their natural environments; for, as Björk ingenuously sings on the song "Crystalline," "with our hearts/ we chisel quartz/ to reach love."

Cultivating uncertainty over irresolvable Anthropocene dilemmas in an ecocomposition classroom may seem at cross-purposes with teaching rhetorical and political-cultural strategies of future-oriented resolve, persistence, and resilience. Yet ecologically aware citizen science paired with sf literature and media expressly intertwines environmental disaster survival narratives and complex identity politics by probing the unruly contact zone between the climatological and the existential, the chemical and the cultural, and the personal and the planetary. Charlie Jane Anders's *The City in the Middle of the Night* (2019) is another representative ecodystopian sf text that embodies this Anthropocene collapse of binaries. Set a thousand years in the future, as humanity struggles to colonize the tidally locked planet January, perpetually divided between sun-facing desert landscape and stars-facing barren nightscape, the novel pits two day-dwelling cities—the authoritarian Xiosphant and the anarchic Argelo—as competing social strategies and political realities for survival "on a planet [that] is antithetical to human life" (73). Just as Le Guin complexly refracted the critical utopian, Cold-War-era cultural politics of Soviet (Anarres) and American (Urras) dynamics onto a world-reduced sf canvas in *The Dispossessed,* Anders transfigures the binary between authoritarian neoliberal and anarcho-libertarian climate-induced critical dystopias. Not only do the human colonists confront the inhospitable nature of this foreign planet, in the vein of Robinson's *Aurora,* but the Anthropocene industrial technology the colonists bring with them has also begun to change the delicate balance of the already fragile atmosphere, leading to rising global temperatures and hazardous storms of acid rain. This dichotomy between the two cities quickly breaks down, then, as Anders introduces the Citizens, a nomadic band of gypsy traders, menders, and mystics who travel between the diametrically opposed cities. Named for their empowered civic agency in this world of bleak politics, the Citizens create personal cosmologies based on their consulting Elementals on the alien planet and "visiting certain dormant volcanoes or other natural formations" (192).

The Elementals the Citizens commune with turn out to be Weird aliens, giant tendrilled, sharp-toothed, furry, lobster-like sentient creatures called the Gelet but known to humans as Crocodiles. The human protagonist of the novel, Sophie, is saved by the Gelet when she is banished into the night from Xiosphant after a minor infraction. Once she communes with the Gelet by putting her head in their mouths, Sophie learns that the Gelet have a hive-mind collective consciousness, and although othered as dumb brutes by humans, these aliens are in fact a technologically sophisticated civilization who has created an underground city in the middle of the night via "networked chains of flora to sequester and redistribute heat energy" (294). After the Gelet surgically transform Sophie into a queered and transgender hybrid of human and Gelet, much like the Oankali-human hybrids in Octavia Butler's *Xenogenesis Trilogy,* Sophie reluctantly becomes a cultural ambassador to Xiosphant and Argelo, which itself has been fused into one overarching society when Sophie's aristocratic frenemy, Bianca, leads an invasion that topples Xiosphant. As a hybrid composite of sustainable alien and destructive human, the novel ends with Sophie just beginning her mission to convince the human colonists to surgically transform themselves and live more harmoniously with the natural environment they inhabit. After all, the Gelet are cyborgian hybrids themselves that blur the fast-and-loose distinctions between nature and culture, the planetary and personal; and Sophie's newfound knowledge stems in part from the Gelet's unique Anthropocenic merging of indigenous science and geoengineering technology that involves "experiments that have been in progress for generations, and processes that move too slowly to observe in one lifetime," and now "since the climate destabilized, their main goal has been to protect future generations" (315). The resilient hope for a technologically sophisticated reconciliation with a troubled and unstable environmental hyperobject here is not a nostalgic call for a romanticized or pastoral return to a primitive oneness with nature; rather, the novel nurtures a determined, hopeful, promissory desire for a dynamic, tentative accord with a transformative, sustainable citizen science.

THREE

Negotiating the Art and Rhetoric of Artificial Intelligence and Citizen Science through Science Fiction

CITIZEN-SCIENCE–AFFILIATED PROJECTS, SERVICES, AND GAMES

Shelley, A.I.
OpenAI's GPT-3
TuringBox
Google Search

SF LITERATURE AND READING

2001: A Space Odyssey, by Arthur C. Clarke
Destination Void, by Frank Herbert
The Einstein Intersection, by Samuel Delany
"When Robot and Crow Saved East St. Louis" and *Autonomous*, by Annalee Newitz
The Corporation Wars, by Ken MacLeod

SF MEDIA AND MULTIMODAL TEXTS

O.K. Computer, by Radiohead
Star Trek: The Next Generation, by Gene Roddenberry
Vision, by Tom King

Raising a wider public awareness about deep-learning artificial intelligence (AI), in 2017, researchers at MIT Media Lab created a state-of-the-art algorithm, trained on amateur, crowd-sourced horror fiction, to collaborate with human writers through online social media in the generation of original horror stories. This project, aptly dubbed *Shelley, A.I*, an homage to Mary Shelley's *Frankenstein* (1818), was widely covered in the news media in tones of mild alarm, absurdist whimsy, and diverting wonder. Yet this (anti)christening may anticipate a deeper and more significant registering of the impact of AI systems soon to reshape our current technocultural imaginary. After all, Mary Shelley's foundational role as proto-sf novelist is ritually credited not only with popularizing existential angst over machine intelligence, but also—in a longstanding interpretation of the canonical novel that has, for instance, recently been revived by Despina Kakoudaki—over the years many have regularly read Victor Frankenstein, qua the archetypal Mad Scientist, to function as an avatar of the lone, defiant scientist-hero whose transgressive meddling in the "unhallowed arts" (Shelley 190) dares to unsettle the entrenched status quo of conventional societal mores. In other words, Dr. Frankenstein represents science as disruption.

Today news stories often report that prodigious advances in automation once again threaten to unsettle the labor market on an untold variety of political-economic fronts. Computer-vision AI, and its newfound ability to selectively plant crops and systematically spray tailored pesticides, threatens to replace farmworkers; voice-operated virtual assistants, such as Apple's Siri or Amazon's Alexa, might soon supplant hotel concierges and other customer-service representatives; and self-driving cars could possibly phase out taxi and truck drivers. Moreover, the ubiquity of already existing AI systems in contemporary everyday life has only begun to be formally studied. Such already deployed AI systems extend their influence in spheres as diverse as micro-targeted advertising, electioneering, job-candidate screening, and spam filters, extending everywhere from medical-care screening to bank loans, from facial recognition and predictive patrols in policing to judicial sentencing to deepfake computer-generated videos. I hasten to add that even though working journalists now compete or collaborate with writing bots such as *The Washington Post*'s Heliograf for their livelihoods, the prospect over the automation-driven replacement of flesh-and-blood human writers with deep-learning neural nets in the workplace or the classroom presents manifold practical technical hurdles to overcome before becoming feasible, let alone inevitable or desirable.

Nevertheless, passable AI-generated natural-speech writing is looming just on the horizon of possibility now more than ever, as evidenced, for instance, by cofounder Elon Musk's OpenAI's GPT-3, which, devised by AI researchers Alec Radford and Ilya Sutskeve, and trained on 175 billion parameters and a web-based corpus of over a trillion words, can produce lengthy, intermittently coherent essays in response to short prompts. Trailing in second place in a stiffly competitive marketplace for natural-speech generators, Microsoft's Turing NLG, based on 17 billion natural-language parameters, likewise is under development to interact with people by fluidly fielding questions, engaging in articulate conversations, and interpreting documents. Near-future natural-language generators will certainly use trillions of parameters and will be vastly more compelling human-voice surrogates, even though many science commentators believe this eventuality only increases the necessity for human-AI partnerships. As Janelle Shane, in *You Look Like a Thing and I Love You* (2019), describes the built-in limitations of algorithms, "Dealing with all the unpredictability of a city street or a human conversation is probably beyond its reach—if it tries, it may succeed much of the time, but there will be glitches" (213). From the ongoing practice of pedagogically evaluating the machine assessment of student writing in college composition to the vexing issue of assessing the role of the human voice in an academic or professional piece of writing at all, this looming specter of formalized writing automation hovers over the future of teaching writing and communication as an academic discipline. In *Robot-Proof,* Joseph A. Aoun, the president of Northeastern University, reasonably predicts that "with machine learning on the rise, it will probably be no great leap for [algorithmic programs] to write A+ essays for themselves sometime soon" (50). Aoun contends that such an inevitable prospect is not a doomsday scenario, however, but rather a transformative opportunity for the discipline.

Exhumed from 140,000 stories posted to the Reddit "r/nosleep" forum and patched together instantaneously with the help of social-media participants, *Shelley, A.I.* was expressly designed to incite collaboration between advanced language-modeling algorithms and everyday citizens online. Likewise, one of the researchers behind the *Shelley, A.I.* project, Iyad Rahwan, also has cowritten a paper called "Closing the AI Knowledge Gap" that proposes the TuringBox, an online forum for AI researchers, who are all too often constrained by intellectual property restrictions, to upload, study, and test AI algorithms. The authors of the paper contend that such information sharing that the TuringBox promises has the "potential to democratize the study of AI behavior." Much akin to the democratizing bent of many citizen-science projects, the TuringBox therefore is a participatory bid to expand the study of artificial intelligence beyond isolated engineering problems and specialized software innovations, bringing AI down from the rarified realm of general scientific inquiry into its real-world, everyday-life, citizen-science applications. As our professional careers and ordinary lives become more deeply integrated with ubiquitous AI systems, such information-sharing projects become vital not only to researchers but also to attempts by concerned uncredentialed lay citizens and nonspecialists to grasp the seeming redundancy of human labor when confronted with this perennially resurgent Frankenstein's monster of disruptive automation.

In this chapter, I argue that sf literature and media, presenting our technologically saturated dilemmas in a gripping and provocative fashion, can provide students with opportunities to explore the best of what is science-fictionally thought and said today. Such pedagogy has the potential to be nondemoralizing and accessible, thus curbing the feelings of dread, anxiety, and desperation that sometimes accompany the rigorous teaching of basic skills. In a reform-driven argument for the increased use of using literature to teach rhetoric in the college classroom, John Briggs, for instance, contends that "paradoxically, it is one of the best ways for specialists to communicate with generalists (the public), and with specialists in other ways" (623). In a time when *The Economist* proclaims that 47 percent of jobs will be automated out of existence in the near future, the general and holistic approach to engaging with the versatile, systems-oriented rhetoric of disseminating specialist and technical knowledge more widely, generally, and holistically is especially attractive for employers interested in the coveted jobs of the future, such as, for instance, the data scientist, an in-demand position which requires professionals to hone deeply reflective stances that discretely analyze, reshuffle, and synthesize the sophisticated science-driven data maelstroms that shape our contemporary reality. Yet to prepare for our intensely automated digital futures in which people will need to outsmart and train algorithms for a living, students in higher education must overcome a default setting of being, if not completely science-illiterate, at least habitually science-allergic and science-phobic, an attitude that is surprisingly pervasive in our technologically saturated global society.

Since the landmark studies conducted by Margaret Mead in 1957 and its follow-up by David Wade Chambers in 1982, it has been generally acknowledged that a widespread hostility to technocratic military-scientific research has deep roots in recent global history. Citizen-science projects, such as those proposed by machine learning researchers above, vie to dismantle this long-standing adversarial attitude toward science and technology without necessarily also disavowing the abiding critique of what John Passmore considers "aristoscience," or the elite, technocratic domain of pure, theory-driven science. According to Passmore, a prevailing antiscience cultural outlook stems from how science is taught and disseminated in contemporary culture, often through the insulated prerogatives of cloistered institutions and hierarchical bureaucracies that generate research of such recondite abstruseness and dissociation as to be deliberately divorced from the everyday life of untrained ordinary people. Against such widespread antiscience beliefs, Passmore develops a spirited defense of what he terms the "pro-science" attitude whose popularization he chiefly equates with scientists themselves, science journalists, and sf writers who serve as "intermediar[ies] between science and the general reader," therefore "enlarge[ing] horizons" about "problems which are of major concern to society" (60). What Passmore calls for here should not be hastily reduced to the common sense of public outreach mounted by scientific institutions historically but a more sustained, critical interrogation and active civic participation in science by engaged citizens themselves, as has only begun to be glimpsed today through recent citizen-science projects, programs, and endeavors.

As discussed in the introduction, James Wynn characterizes the rhetoric of citizen science more broadly as an attempt to not only reverse the rising tide of scientific literacy but of educating and promoting deep identification with scientific debates and discourse. Hence Wynn describes citizen-science rhetoric as not interested in cultivating trivial knowledge or the rote accumulation of irrelevant facts by the otherwise uneducated or inexpert; rather the participation and engagement of interactional and contributory citizen science consists of the emergent phenomenon in which laypeople assume a civic responsibility to contribute to scientific enterprises relevant to their everyday lives. Such citizen scientists seriously contribute to scientific discoveries and technological inventions by not only facilitating meaningfully in the process of gathering crowd-sourced data as virtual field assistants but also in the far more integral role of generating research questions, testing hypotheses, assessing risks, demystifying expertise, and even making policy, as each section of this chapter will respectively discuss in turn. Such a public enterprise—a quasi-utopian dream of democratized science—would disavow C. S. Pierce's pragmatist claim that "scientists are the best of men" in favor of Passmore's more democratized notion that "progress does not originate in the laboratories" so long that "science is not left to the scientists" (60).

For, as Kelly Moore has demonstrated, the historical roots of the public disenchantment with the militaristic, authoritarian impulses that Max Weber famously critiqued as "science as vocation" and Herbert Marcuse lambasted as "instrumental rationality" trace back to the ideological implication that scientists are simply not like the rest of us commoners, that is, the laypeople or general public. The exigencies of funding needs and political pressures of scientific professions have led many to associate so-called Big Science with the agendas of regulatory state apparatuses and corporatized capitalist markets. Moore documents the red-baiting of the 1940s and 1950s that succeeded World War II and the Manhattan Project, which led to the dominant Cold War defensive posturing of twentieth-century aristoscience when the Anglo-American scientific establishment unconvincingly but insistently claimed to be apolitical. Against such centralized secrecy and elitism, vociferous antiwar protests, antinuclear advocacy, and civil-rights agitation of the 1960s and 1970s pitched their dissident political campaigns of activism and advocacy. Yet the speciously apolitical defensive posture only entrenched the asymmetrical current norm in global scientific institutions. Indeed, a disruptive countertrend of subcultural politicized science is still only barely visible today, as exhibited in the recent global March for Science on Earth Day, April 22, in 2017 that galvanized tens of thousands of participants around the world under repeated attack by the Trump administration's antiscience, posttruth policies.

This chapter joins in spirit with projects like *Shelley, A.I.* and TuringBox to probe the value of acclimating a vibrant citizenry, especially in the secondary and higher-educational environment, to sf literature, media, and art precisely for the purpose of limning the productive intersection between the rhetorical, cultural, and social understandings of AI. What counts as an AI algorithm here is deliberately ambiguous, defined loosely as the products of a highly embattled and constantly evolving arena of scientific and technological exploration. This chapter will therefore analyze the ways some representative sf texts can facilitate critical interrogation of recent, future, and past advances in machine learning both inside and outside the classroom by modeling how citizens can collaborate meaningfully in the process of data-gathering, hypothesis-testing, problem-solving, and policy making. Over the past century, a rich, complex body of sf literature and media has evolved toward the direction of citizen science not despite but precisely because of the fabled rise of thinking machines. Once what Darsie Bowden has called the "mythology of voice" becomes increasingly subjected to intense scrutiny by the entwined advances of contemporary linguistics and AI systems, it is therefore not only likely but a probable certainty that in the near future, a computer will be able to produce a piece of writing containing all the subtle and elaborate credibility of a human writer. Correspondingly, a primary question of this chapter is the following: what is the role of literacy, composition, science, and citizenship education given such a long-vaunted singularity will eventually become a dominant reality?

Natural language generation is a major subfield of AI research, and the current results are mixed but promising, as exemplified by the fascinating children stories generators such as those created by the AI Scheherazade, developed by Mark Riedl and colleagues at the Georgia Institute of Technology in Atlanta, or the Shakespearean sonnet generators such as those created by Jey Han Lau at IBM Research Australia, or Botnik's Harry Potter novel-generator that produced the whimsical *Harry Potter and What Looks Like a Portrait of Ash*. The current state of the art does not seem impossibly far off from how Richard Powers in the mid-1990s encapsulated the language-generation field in his postmodern novel *Galatea 2.2* (1995), a heartbreaking work of fiction that ultimately casts serious doubts over whether an AI system will ever be proficient enough to write an undergraduate essay that analyzes literature indistinguishably from a gifted English major. Yet that such a machine intelligence could succeed even partially entails a "cheat so colossal and magnificent that it takes your breath away" (Powers 245). This rhetoric of wonder and applicability attached to science popularization explains why AI natural language generators garner such exaggerated media attention since, as Jeanne Fahnestock argues in "Accommodating Science," such journalistic accommodation revolves around the Aristotelian epideictic appeal to "add to the significance of subject by claiming its uniqueness, its one-of-a-kind status" (280).

A vexing problem with simulating a human writer is the issue of voice, which pedagogical linguists and composition theorists are still endeavoring to quantify more precisely. Voice might be classified more technically as the probabilistic occurrence of certain linguistic cues, such as, for instance, the relative salience of the easily recognizable voice markers of intensifiers, hedges, boosters, directives, questions, attitude indicators, and shared knowledge appeals (Zhao 73). It is certainly conceivable that a next-generation AI program might be trained to simulate such voice markers believably; yet the mythology of the human voice in writing will likely still remain an unfinished project despite such coups. Once hard technical problems such as simulating voice markers are mastered, it is easy to speculate that the need for training human writing in the workplace and in education will mutate into a less technically quantifiable factor. For instance, in one plausible near-future pedagogical scenario, the education of young writers will depend on training citizens to think critically and deeply about how to better design the next wave of AI systems and devices. In the domain of natural-language generation, for instance, these next-generation AI systems will also need to more closely approximate the singularity of the human voice conceived as an ever-receding horizon of possibility. Such a pedagogy would optimize full and active participation in the construction of beneficial scientific knowledge for the everyday lives of citizens, complete with all the checks and balances that such critical participation entails. This civic-minded approach to future composition education will constitute an obligatory survival skill in the AI-dominated digital future. Long enmeshed with anticipating a future transformed by AI, such pedagogy may find inspiration, encouragement, and insight by reflecting on the untapped reserves of sf literature and media in the classroom.

"I'M SORRY, DAVE, I'M AFRAID I CAN'T DO THAT": GENERATING THOUGHTFUL INQUIRY AND REFLECTION ON THE POWERS AND LIMITS OF AI RESEARCH

Arthur C. Clarke became a world-famous household name for his futuristic, fictional extrapolations on AI and space travel; much like the more recent projects of *Shelley A.I.* and TuringBox demonstrate as a citizen-oriented intervention into the public conversation on AI systems, Clarke's fictional works, for all their visionary technologically optimistic exuberance, most often underscore the pitfalls and problems inherent in the synthesis of massively data-crunching, recursively sophisticated automation algorithms, and their influence on intricate networks of entrenched scientific systems, technologies, and political-cultural ideologies. The AI systems that Clarke frequently imagines are flawed creations, capable of great leaps of posthuman nuance and sensitivity, and yet fundamentally conforming to contingent historical actors who cloak their politicized agendas and interests under the guise of scientific rationality. Such rhetoric can indeed seem tantamount to rhetorical appeals to the democratizing and participatory impulses of citizen science, as when the *Shelley A.I.* designers endeavor to circumvent intellectual property copyright that hamstrings open-source scientific research. Such rhetoric can also be more disruptively beholden to narrow capitalist imperatives, as when in our own time the company Narrative Science develops a nominally AI neural network algorithm called Quill, which touts itself as a natural language generator and which, as its marketing brochure proclaims, can be "taught automatically to analyze, interpret, and communicate insights from your data in the tone, style, and language of your business."

Ironically, the computer processing of a so-called "deep-learning neural network" invokes a computer-brain analogy to enshrine its own putative neutrality and inevitability; in contrast, critics contend that a so-called neural network can be considered learning only in a cursory sense given that no conceptual mapping or underlying complex understanding of the world underpins, suffuses, or precedes its cognition. Instead of organic cognitive learning per se, a neural network tweaks knobs of weighted layers, and based on a series of outputs and inputs self-adjusts complex nests of self-reinforcing computer code, simulating the behavior of neurons only in the narrow sense that neurons analogously form patterns of connections based on a combination of external sensation and internal brain input. The neural network indeed may be categorized as a compressed form of classical rhetorical scheme known as "antimetabole," which, as Jeanne Fahnestock explains in *Rhetorical Figures in Science*, may be "epitomized by joining the two separated halves . . . 'we treat the organism as a complex machine; machines are coming to resemble biology'" (Fahnestock 41). This rhetorical figure that links mind and machine has been heavily fostered by successive waves of research discourse on AI systems that underlies much of the checkered cul-de-sacs as well as the stunning breakthroughs in the field's long history. Paul Edwards views an awareness of the inherent limitations of AI neural-network rhetoric as constitutive of Clarke's representation of closed, hyperrational supercomputers tragically constrained by all-too-human parameters (323).

Indeed, Clarke clearly recognizes that reducing human knowledge to informational or cryptographical data can be an absurd exercise in hubristic overreach. For instance, Clarke's short story "Steam-Powered

Word Processor" (1986) is a whimsical pseudo-Victorian pastiche written as a mock-forward and -afterward to *Sermons in Steam,* a rare antique book attributed to the small-parish vicar, the Reverend Charles Cabbage of the St. Simians Parish in Far Tottering, Sussex, an obvious satirical send-up of Charles Babbage, who theorized the early computing concept of the difference engine. The rumor that the opinionated narrator of the story wishes to dispel involves the idea that the book of sermons in question was in fact written by the Word Loom, a gigantic retrofitted church organ, that would "automatically reassemble masses of texts . . . [and] create any number of sermons from the same basic material" (Clarke 931). The acerbic humor of the story derives from Cabbage's heretical, vainglorious obsessiveness as well as its clash between the engineering blunder the elite scientist attempts and the provincial small-mindedness of rural England. Cabbage's folly is that he refuses to countenance the unmanageable technical challenges of the voice-simulating feat he ultimately fails to accomplish; Cabbage, for instance, must harvest the steam power of an industrial threshing machine to prepare a single sermon in ten weeks, and even that proves to be a ridiculous pipe dream; for once he removes the stain glass windows from the church to install the intricate machinery of pumps, tubes, valves, and belts for his steampunk automatic language generator, the Archbishop of Canterbury resorts to exorcism and reconsecration of the parish. In a bathetic ending, Cabbage then elopes to Australia with the village blacksmith.

In case the literary provenance of Cabbage's delusions of grandeur remain undetected, Clarke makes the influence explicit when he mentions that the latest living descendent, Miss Drusilla Wollstonecraft Cabbage—and the implied allusion to Mary Wollstonecraft Shelly—does not approve of the narrator's glib dismissal of her illustrious Victorian ancestor. It seems, though, that the maverick scientist opened himself up to the ridicule of future generations as soon as he treated the complexity of sacred writing as the endlessly iterable data that could be reassembled as if by mindless monkeys on typewriters. Yet the epic failure does not foreclose the possibility of eventual success in our own future, as even the otherwise mordant narrator speculates perhaps a single page of the book, rife with typographical mistakes, was the "most remarkable—and misguided—technological effort of the Victorian Age" (934). Cabbage, ethically dubious and steeped in egocentric pretensions as he may be, also provokes such derisive mockery because his recklessness and disruption challenge the quiet traditions of rural provincial life, even as he recruits skeptical citizens to pursue his outrageous agenda.

As Clarke repeatedly imagines it, the dangers with the panoptic eye of artificial intelligence dwells in both its keen perceptions of and its myopic blindness to empirical knowledge transformed into symbolic computer data. In the novelization of *2001: A Space Odyssey,* Clarke makes explicit the essential narrative argument of the celebrated film that the Cold

War–era governmental secrecy to keep members of the starship *Jupiter* out of loop about their investigation of an alien monolith signal in fact initiates the fatal malfunction of the AI ship computer, HAL (Heuristic Algorithm) 9000, who then, naturally enough, attempts to murder the hibernating crew out of an all-too-human instinct for self-preservation. This sequence of information-processing error begins when HAL first makes a false prediction that the AE-35 unit that controls a communications antenna signal will inevitably fail, only to be threatened with termination himself after this prediction is proven wrong by the astronauts Dave Bowman and Frank Pool, when they duly replace the unit and send the diagnostics back to Mission Control for analysis. In the novel *2001,* Clarke both invites and militates against the anthropomorphizing of the neural network of machine intelligence when he writes, "deliberate error was unthinkable. Even the concealment of truth filled him with a sense of imperfection, of wrongness—of what, in a human being, would have been called guilt" (148). Data, as contemporary readers are well aware, does not need to appear veridical nor trustworthy to be electronically stored, analyzed, and manipulated in a computer algorithm, and the introduction of the anthropomorphized guilt-equivalent error into the HAL 9000 system is a direct unintended result of its problematic programming, and its adherence to the exclusionary, rote rivalries of military-industrial nation-states and the techno-imaginary of elite corporate powers.

In the sequel *2010: Odyssey Two,* HAL is rehabilitated when investigators discover its murder spree was rooted in the inexorable logic of faulty programming: "The situation conflicted with the purpose for which HAL had been designed—the accurate processing of information without distortion of concealment" (Clarke 155). Confounded by the social and political priorities that shape and predetermine its transparent processing of data, HAL develops what a computer-scientist troubleshooter calls both a human-like "schizophrenia" and a machinic dilemma of being "trapped in a Hofstadter-Mobius Loop." The protagonist Heywood Floyd immediately casts doubt on this diagnosis, however, given the "defensive" (156) postures of the programmers, ultimately relaying to the National Council on Aeronautics that a report in "non-technical terms for the Council" (153) would be impossible, thus reinforcing technocratic hierarchy. And the outward thrust of the *2001* series as a single serialized thought experiment complicates this dilemma even more densely when it is revealed that aliens who deposited the monolith have "invaded" HAL as a "ghost in the machine" (247) before evolving into a Star-Child boon companion to his erstwhile human lobotomizer Dave Bowman. Even though the algorithmic computation of data may lead to stunning leaps in knowledge that the best scientific theory cannot adequately frame, such computation is also constrained by the confirmation biases of its original research questions and agendas.

The implications of HAL's extraterrestrial co-optation are twofold. Firstly, HAL's coldly logical instrumental neutrality has become uncannily attuned to data-gathering beyond the narrow limits not only of human programming but even of the Enlightenment project of scientific methodology itself. Secondly, this baffling but workable empirical adequacy originates in the sufficiently advanced magic circle of inexplicable computer algorithms, as later confirmed when the thrust of the series reveals that monoliths are themselves not shepherded directly by omniscient alien superbeings but only the forgotten remnants of a long-since dematerialized civilization that has seeded the universe with "von Neumann machines" (297), themselves self-replicating AI systems that were originally designed to uplift and bootstrap promising species. Von Neumann machines, after all, are any technically programmable machines with internal storage and high-speed memory, whether hypothetically alien or all-too-human in origin. The impetus for HAL's initial meltdown may in fact be connected to his receiving signals from this alien computer and its outpost on the Galilean moon of Europa. The inherent deficiencies of such overreliance on technocratic expertise circle back into narrative frame finally when it is discovered in *3001: The Final Odyssey* that these artificially intelligent monoliths themselves are apocalyptically malfunctioning and need to be deactivated. Nevertheless, as the heroically emancipatory trajectory of a revolution-centered novel like *The City and the Stars* (1956) makes clear, Clarke views the unsettling advancements of AI systems as optimal stepping-stones to guide and safeguard civic-minded interest beyond the stagnant morass of the technologically driven status quo even if, as his story "Dial F for Frankenstein" (1965) about the revolutionary chaotic emergence of a self-aware proto-internet out of the telephone grid also stresses, transgressive milestones may prove disastrous for the entrenched scientific disciplines and inflexible regimes of existing expert authority.

HACKING THE UNHALLOWED ARTS: PROBLEM-SOLVING, TROUBLESHOOTING, AND ENGINEERING PROTOCOLS IN AI RHETORIC AND RESEARCH

The social imperatives that structure, shape, and control the cutting-edge technical apparatus of machine intelligence frequently deploy the rhetoric of scientific breakthroughs, discoveries, genius, and achievement to simplify the messy, doubt-laden problematics of scientific processes and disciplinary debates. Yet the extensive sf literature that explores AI research also posits that radical technocultural paradigm shifts have the potential to overthrow the discriminatory hierarchies and exclusions of elite, pure science. The epigraph of Frank Herbert's *Destination Void* (1966) is to Victor Frankenstein's meddling in the "unhallowed arts" that can "mock the stupendous mechanism of the Creator" (v), a quotation particular to the less heretical 1831 revised edition of Mary Shelley's *Frankenstein*. This "Frankenstein Myth," as Isaac Asimov, author of the influential *I, Robot* series, disparagingly labeled the enduring legacy of Shelly's novel, couples the long-standing culture of popular skeptical resistance and hostility to aristoscience with an awe-inspired appreciation of the techno-sublime of AI research whose internal complexity at times seems to rival that of the vast physical cosmos, or at least the singularity of the human brain itself. Herbert's novel concerns the preplanned sabotage of an interstellar generation starship *Earthling* putatively destined to colonize the star system of Tau Ceti; in fact, the Project Council of Moonbase Control has carefully constructed the emergency scenario of the breakdown of their ship computers, the Organic Mental Cores (OMC), a transplanted human brain, so as to necessitate the invention of a rogue artificial consciousness by an Umbilicus Crew of six trained Doppelgängers, or experimental clones that belong to the scientific project underway. Through recurring chapter epigraphs as well as the general gothic horror of the novel, the making of the sentient, autonomous AI systems is repeatedly compared to the unholy animation of Frankenstein's monster.

The ramshackle invention of a renegade AI in *Destination Void* is, nevertheless, not due to some valorized stroke of single-minded genius, but the accident-prone, troubleshooting trial and error of a team of fringe, blue-sky engineers. The exploratory problem-solving of the clones functioning themselves as guinea pigs in the novel literalizes the robust experimentalism of scientific methodology, involving, as such a verifiability does, hypothesis-testing, indirect validation of predictions, and the gradual elimination of alternative or "control" explanations. Indeed, throughout history the scientific worldview itself has traditionally been viewed as an experimental offshoot of the broad goal of creating machine intelligence. For instance, Enlightenment philosophers constructed the clockwork metaphor for explicating nature, and the inner workings of this metaphor lie at the core of the so-called "Scientific Revolution," which repeatedly invoked the idea of impersonal machine intelligence to justify its legitimacy, as Steven Shapin shows: "Machines might seem *like* purposive agents and even substitute for purposive human labor, and that likeness constituted part of their explanatory power" (Shapin 39, emphasis in original). Updated to respond to 1960s-era computer science, Herbert's hard-sf novel serves as such a valuable pedagogical resource precisely because it vividly illustrates the clockwork purposiveness without purpose of scientific methodology and experimentalism.

In the Enlightenment worldview of modern science, the eventual calibration of consistent empirical-rational knowledge is neither overly theorized nor entirely theoryless. As the brilliantly creative but borderline psychotic clone, John Bickle, demurring *a priori* definitions of consciousness and preferring to examine the construction of an artificial mind as a concrete heuristic phenomenon, bluntly puts it, "We're going to be juggling one hell of a lot of unknowns. The best approach to that kind of job is the engineering one: if it works, that's the answer" (Herbert 49). The fictive future science that the characters rely on to develop the AI system involves troubleshooting indeterminacy, as evidenced by Herbert's trademark world-building, research-driven references to "mocking up prototypes" (58), "built-in misfunction" (91), "roulette cycles" (94), "buffer systems" (106), "diagnostic routines" (122), "homeostats" (138), "neuron blocks" (143), and so on. The active engagement of citizen scientists in AI research can likewise highlight the tentative, haphazard, and inchoate nature of the scientific process in action, especially at the intersection where Big Data meets neural networks and machine learning. It is instructive to be reminded what another clone crew member in the novel, Gerrill Timberlake, realizes: if the data that AI systems mine are the products of symbolic language, then "the symbols are loaded with errors, weaknesses, and flaws" (140).

The experimental probing of error through the testing and refutation of alternative hypotheses reinforces the ad-hoc, algorithmic goal of simulating a seat of consciousness in the novel. Bickle explains to skeptical team members, "If the machine is going to be conscious, we cannot predict its behavior, by the very nature of consciousness, by definition" (150). Herbert draws on Shelley's *Frankenstein* to establish the precedent of such a bewildered machine intelligence, quoting as a chapter epigraph when Adam first speaks, "No distinct ideas occupied my mind; all was confused" (167). The Frankenstein allusions underscore the threat that the AI will go "rogue," that is, as Bickle muses, "we don't know if what we will get will be a useful thing or a monster—the tool or the Golem" (257). At the end of the novel, these anxieties seem partially realized when, after a "white-box transfer" in which Bickle's disembodied consciousness merges with the emergent sentience of the AI system, it takes over the starship and makes preparations for the terraforming a habitable planet in Tau Ceti system for the colonists to land but also delivers this chillingly ironic message to the crew: "You must be together when you make your decision . . . you must decide how you will WorShip Me" (273). Herbert and his collaborator Bill Ransom develop this conflation of theocracy and AI systems—what Ian Bogost in our contemporary cybernetic context has called the "the cathedral of computation"—in the sequels *The Jesus Incident* and *The Lazarus Effect*; and an analogous resistance to subservient devotion to renegade AI as godlike entities forms the deep background of the *Dune*-verse and its Butlerian Jihad, which outlaws thinking machines. Portraying the singularity as tantamount to the ascendency of a new religion, and all the fear and trembling that such a sublime awe entails—replicated today online in Extropian conspiracy theories such as Roko's Basilisk—cements the creation of machine intelligence as a science experiment that could easily run amok. Such allegorical fables thereby reiterate in lurid detail the importance for affected parties of citizen scientists to weigh in when technical experts design experiments.

MINOTAURS, PARANOID ANDROIDS, AND PHILOSOPHICAL ZOMBIES: DOUBT-LADEN RISK ASSESSMENT IN AI RESEARCH

A contemporaneous novella to *Destination Void*, Samuel Delany's *The Einstein Intersection* (1967) imagines the remnants of an AI system discarded by a mysteriously dematerialized human species. This machine intelligence is a minotaur-like PHAEDRA (Psychic Harmony Entanglements and Deranged Response Association), housed in a labyrinthine underworld on a largely depopulated Earth; a postapocalyptic relic, PHAEDRA stores the last remaining cultural memories pertaining to the entirety of the vanished human civilization following its seminal achievement of a techno-rapture into a dimly understood transcendent alternate dimension. In explaining the forgotten future history to the saga's heroic protagonist, Lo Lobey, another mutant scavenger, Spider, suggests the fantastic future science of the enigmatically absconded humanity had its roots in the twin conceptual breakthroughs of Albert Einstein's counterintuitive Theory of Relativity and Kurt Gödel's equally mind-bending Incompleteness Theorem. In the popular context of his pulp-influenced sf novel, Delany is mythopoetically riffing here on the well-documented historical friendship between Gödel and Einstein, as well as the actual influence of Gödel on Alan Turing and the early genesis of computer science. For Alan Turing studied Gödel to devise his idea of the "universal" machine—that is, a general-purpose device that could be programmed to perform an algorithm, or an illimitable sequence of operations according to any given set of formal instructions. Another nested allusion to the history of science hovering in the background is Einstein's famed controversial rejection of Werner Heisenberg's Indeterminacy Principle, hence the "entanglements" in PHAEDRA's acronym.

Yet at least in this fictional world, God does play dice with the universe, as illustrated by the triumph of a Gödel-influenced advanced computer science. PHAEDRA constructs a binary between the famously shock-haired theoretical physicist—whose rigorous, paradigm-shifting notions of physical relativity posited that position, mass, and velocity of an observer affected the uniformity of gravity and time—and the influential formal mathematician-logician: "Einstein defined the extent of the rational. Gödel stuck a pin into the irrational and fixed it to the wall of the universe so it held still long enough for people to know it was there" (128). The implication is that the undecidable algorithms of this future computer science form the basis of computation that challenges the rational validity of the human understanding of the universe itself. This indeterminacy behaves in a fashion distinctively unlike what abstract mathematics can formalize, which corroborates how Roger Penrose interprets Gödel's contributions to science and that many mathematicians still believe—namely, the mind has the potential to intuit formal, Platonic *a priori* mathematical truths an AI system could never feign to emulate. Through this dense tissue of titular allusions, Delany suggests that the techno-sublime AI systems of the future will perform cultural work seriously in danger of being adopted with an uncritical, unquestioning faith once extricated from its immediate historical context. Much later, at the dawn of the increasingly technocratic digital era in the mid-1990s, the remarkably consistent Delany reiterates, "On a material level, our technology is becoming more like magic, with a class of people who know the incredibly complex spells and incantations needed to get the stuff to work, but almost none of whom can get in there and fix it" (192).

By framing the scientific pursuits at the dawn of computer science as a quixotic quest that increases indeterminacy, Delany's novella raises a fascinating dilemma vis-à-vis the problematic role of citizen scientists in evaluating expertise. Doubt, distrust, skepticism, and questioning credibility are certainly cornerstones of institutional science safeguarded by such means as double-blind peer review and committee-vetted allocation of grants; however, the propagation of elite science does not equally reward the generation of any and all research hypotheses as equally valid, and the systemic validation of scientific experimentation from laypeople and nonexperts is not awarded comparable incentives as the contributions of professional scientists. In the cases of AIDs research and air pollution, for instance, Jason Coburn cogently argues that the robust activism of citizen science was galvanized precisely by interrogating the automatically presumed trustworthiness that the public normally confers on institutional aristoscience (66). Delany's interest in unnamable slippages between what is considered official science and deemed legitimately rational over against groundswells of what is popularly deemed irrational, pathological, or illusory becomes a recurring motif in the novella's complex treatment of mythological quest-narratives. As another scholarly character in the novel, La Dire, states—explaining the palimpsestic polygenesis and self-inconsistency of oral storytelling that cheekily maps the hysteria-inciting Orphean lyre onto 1960s-era Beatlemania—early on in the novella: "We've had quite a time assuming the rationale of the world. The irrationale presents just as much of a problem . . . in myths things always turn to their opposites as one version supersedes the next" (17–18).

In one of the interposed author's diary extracts that function as chapter epigraphs to the novel, Delany writes, "The central subject of the book is myth" (78). Clearly, in addition to referring to the dynamic origins of ancient fantasy storytelling, Delany uses the term in the sense that Roland Barthes, in *Mythologies* (1957), also deploys: namely, an ideological system of self-referential, interlocking cultural norms and values. The overdetermined dream logic of the novella underscores that challenging technocratic scientific authorities is paradoxically vital to the visionary dream of scientific progress itself. Such an outlook of citizen-science skepticism interrogates the default status quo of kowtowed deference and inferiority, in which technical devices and institutional regimes of technocratic objects in the world dictate the operations of networked human agents and individuals in a process that Bruno Latour calls "prescription" (Latour 157). This skepticism, of course, does not seek to invalidate scientific worldviews categorically but rather emphasizes the rigorously skeptical approach of scientific refutability itself that treats scientific knowledge as a continually evolving social and cultural product and which supplements and qualifies grandiose claims to neutral, transhistorical rationality. Science as cultural mythology reworks the empirical frame of reference not as univocally legitimate but as one of many decentered perspectives.

That Delany's protagonist, Lo Lobey, encounters the Minotaur of AI systems in an underground labyrinth is, of course, no coincidence. In her historical study of behaviorism and its extensive use of rat mazes, Rebecca Lemov calls our attention to think critically about the semantic similarity between labyrinths and laboratories: "The maze had long stood for the struggle to find one's way when truth was elusive and the way fraught with monsters and despair: Theseus had killed the Minotaur in the labyrinth; the Christian Wayfarer of *Pilgrim's Progress* sought God in one; and Nietzsche begged to be lost in one" (Lemov 37). From the post-apocalyptic vantage, the disoriented Lo Lobey must navigate the haphazard evidence and cryptic artifacts gathered from what Delany, in another one of his journal-extract chapter epigraphs, frames as the prototypical "modern" (Delany 79) condition, namely, the Orphean backward gaze at the labyrinthine laboratories of humanity's discarded history, as if from the dissociated perspective of an obsolete AI system. In Delany's novella, Kid Death colludes to enlist Spider in patricide of his equally murderous father with this plea: "Life and death, the real and irrational aren't the same for the poor race who willed us into this world" (85). The cultural imagination of AI systems, after all, envisions the singularity as the epitome of human rationality, ingenuity, and complexity for a well-meaning, authoritative reason. For instance, since Paul Meehl's influential studies in the 1950s, AI research has regularly verified that mathematical algorithms have consistently outperformed human decision-making with a near-perfect track record. Yet Delany's literally posthuman fantastic future radically reimagines the basic historically situated underpinnings of such algorithms so that grand claims to universal rationality are rendered obsolete and perfunctory, tragically backwards-looking like Orpheus journeying empty-handed from the underworld. Delany anticipates that the conceptual breakthroughs, radical discoveries, and paradigm shifts of future societies will provide revolutionary new frameworks of rationality and irrationality, extraordinary new research questions, and prodigious new systems of scientific doubt. In this brave new world that still hews closely to the essential principles of scientific methodology, the Olympian AI systems of the past will necessarily be viewed as benighted and discardable waste.

Naturally, the myriad risk factors of AI algorithms that fall under the broad canopy of machine learning should determine the relative degree of public concern and citizen engagement. Whether one's digital entertainment streaming ques are limited, or whether one's bread and breakfast vacation lodging experience is suboptimal, are mostly trivial, consumer-oriented quandaries, if still symptomatic of larger systemic problems. However, the cumulative net effect of AI-induced automation may be, as Nicholas Carr argues in *The Glass Cage,* deleterious in numberless ways beyond the significant one of a loss of gainful employment for legions. For automation "narrows our perspectives and limits our choices" (Carr 2). In particular, as Hannah Fry argues in *Hello World,* when a random-forest algorithm identifies a false positive in the denial of bail or parole in a court's sentencing decision, or when neural-net algorithm leads to a false negative in the diagnosis of a disease, or when a teacher is fired or hired based on an inscrutable algorithmic calculation, these are failures that cannot be chalked up as regrettable collateral damage.

Hence Alec Ross posits that the many vocations not automated into oblivion in the near future will need to negotiate, manage, and supervise machine intelligence "correlations made by big data are likely to reinforce negative bias . . . [and] because big data often relies on historical data or at least the status quo, it can easily reproduce discrimination against disadvantaged racial and ethnic minorities" (Ross 184). In particular, in *Weapons of Math Destruction,* Cathy O'Neil underscores the dangers of predictive policing, aggressive advertising algorithms by for-profit vocational schools, corporate and governmental surveillance regimes, and insurance companies denying coverage as reasons why more redundancy pathways, checks and balances, and backup safeguards are needed when devising algorithms. Bottom-up citizen advisory panels, grassroots nongovernmental organizations, and citizen calls for market oversight and regulation need to anticipate potentially catastrophic failure cascades induced by poorly understood and hastily implemented AI systems.

A chief high-risk factor for AI research is the professional and pedagogical impact of automation on higher education. An associated risk is the impact of these sophisticated programs on the cultural politics of civil discourse. If training students to compete with algorithms will amount to a necessary survival skill for their digitally dominated future, training students to negotiate and challenge algorithms will be equally vital to the longevity and resilience of public communities in the future. As discussed earlier in the chapter, the neural-net metaphor of these programs builds on a long history of AI researchers making more or less grandiose claims about emulating human intelligence. Nevertheless, perhaps it is wiser to follow Neal Stephenson who prefers in *The Diamond Age* (1995) to call the phenomena "pseudo-intelligence" (22) as a more apt descriptor of computer algorithms, since such "neural nets" in fact merely consist of remarkably efficient, if sometimes dimly understood, layers of goal-oriented automated self-reinforcement programming. This code is then labelled "intelligence" and "learning" for the sake of intellectual convenience. The high risk of failure in AI systems justifies the growing public panic over such putatively neutral tools of machine intelligence. Hence, as Siva Vaidhyanathan highlights in *Antisocial Media,* given that a psychographic company such as Cambridge Analytica that used advanced AI algorithms to micro-target political advertisements on social-media sites may have had a nonnegligible influence on a close, divisive election, the sheer uncertainty over the potential erosion of democratic institutions in the future motivates the need for swift and bold public intervention and more concerned civic discourse. Indeed, Vaidhyanathan argues that the information pollution of media-savvy, AI-dominated political engineering undercuts the sovereignty of civic duties: "Citizenship grows 'thin,' as too much information lies within convenient reach but is cacophonous, confusing, and contradictory" (164).

Impassioned civic discussion in a writing classroom may be sparked by assigning popular music that addresses problems of automation, digital surveillance, electioneering, virtual reality, and machine intelligence. For instance, sampling songs and polling students to see whether Radiohead's classic album *O.K. Computer* (1997) still speaks to a postmillennial generation often popularly characterized as being addicted to social media and smartphones can be energizing for a classroom lapsing into textual doldrums. Created during the ascendancy of personal computers, a song from the album "Fitter Happier" uses the inflectionless synthesized voice of a MacIntosh SimpleText speech system called Fred to drone a neoliberal ethos of self-improvement, well-being, conformism, and personal growth from the perspective of a terminally depressed associate employee of some faceless corporation. The two-minute harangue of adjective and noun fragments accompanied by a lugubrious musical arrangement—the litany begins, "fitter, happier, more productive/ comfortable (not drinking too much)/ regular exercise at the gym (3 days a week) . . . "—derives its humor and pathos from the eerie sounds of the automated voice rehashing bland human psychobabble, health-industry bromides, self-help mantras, and corporate speak. In other words, the track highlights the problem of preserving the mythology of an authentic human voice in a landscape dominated by digitally mediated communication; from the start teeming with anxiety and ennui, the track becomes increasingly unhinged, ending with the bleak return to the "fitter happier more productive" descriptors before ending with a tragic act of identification with "a pig/ in a cage/ on antibiotics." This line recalls a snarling insult hurled earlier in the album, on the song "Paranoid Android," which lacerates a "kicking squealing Gucci little piggy" encountered in a barroom brawl.

The reference to a paranoid android is itself an homage to Marvin the depressed AI robot from Douglas Adams's *The Hitchhiker's Guide to the Galaxy* (1980). The MacIntosh synthesized voice also recurs on the track "Paranoid Android," claiming the voice is indeed "paranoid" but not an "android," resenting the Phildickian dehumanizing epithet. The anxiety-ridden, self-loathing persecution complex of this voice is deep in denial, primed like a prize pig for optimal human consumption, fated to be a sacrificial offering, despite all its earnest effusion of superficial buzzwords and pop psychology. The android voice that haunts the album likewise scapegoats and violently expels with petty, vindictive righteousness, contempt, and schadenfreude all those it finds personally distasteful and disappointing. It appears that the maladjusted android might even subscribe to the Big Brother biopolitical policing found in another song on the album, "Karma Police," rehearsing barrages of empty platitudes and truisms, "buzz[ing] like a fridge . . . like a detuned radio." Raised on viral memes, scroll-trances, cyberbullying, tweetstorms, and oversharing, students today may find profound affinities with how *O.K. Computer* as a whole captures the postindustrial malaise of contemporary life, the self-quarantined angst of being devoured by the all-consuming Moloch of our virtual lives, the wired quiet desperation of being constantly tethered to always-on devices both personally and professionally.

The album may also strike chords in musically inclined students as the joyless simulacra and anomie of being a copacetic cog in the machine—that is, morphing into an alienated android voice—are contrasted with the awe-inspiring power of the intricate sonic compositions. The angelic pitch and quavering timbre of Thom Yorke's vocals lyrically express a cathartic introspection. Jonny Greenwood's minor-key chord progressions and keyboard melodic hooks complement this lyricism with an all-encompassing lachrymose mood. Then Colin Greenwood's insistent bass-guitar rhythms and Philip Selway's loud and fast stadium-rock drum fills finally hammer home a disgruntled disaffection. Likewise, Ed O'Brian's layered electronic effects, ambient noises, woozy reverb, whirring delays, wobbling synth filters, and the other assorted digitized distortions concocted by the band, evoke fleeting moments of transcendence in a static wasteland of airport terminals, suburban enclaves, and busy freeways devoid of genuine human affect or value. All this rich aural material provides ample opportunity for students to write analytic and evaluative arguments on the endlessly fascinating artistic semiotics of the album, and either its resonance or datedness for their contemporary generational concerns and attitudes.

To understand the unexpected fallout of AI algorithms on the social-media universe, such as those feared to be unleashed by Open AI algorithms, it may be worthwhile to consult Charles Perrow's *Normal Accidents* that suggests that discussions of the complex, systemic risk factors of scientific projects such as the building of nuclear power plants—either the "pure" rationality of actuarial cost-benefit analysis or the "bounded" rationality of cognitive psychologists—too often berate and dismiss the intuition of citizen contributions to scientific decision-making and policy, neglecting that public concerns helpfully revolve around "social rationality"—that is: "The public is uninformed in many respects, and certainly can make errors in reasoning, but for matters of catastrophic risk these errors seem less disabling than the alternative of neglecting the public rationality embedded in social and cultural values" (324). Hence those who minimize the risks of AI systems on society reinforce a dangerous, damaging power dynamic that is incentivized to overlook negative impact, especially as such risks adhere to minorities in majoritarian democracies. For example, in *Algorithms of Oppression,* in a discussion of the structural racism inherent in the proprietary Google search-engine technology, Safiya Omaja Noble reminds readers that "deep machine learning, which is using algorithms to replicate human thinking, is predicated on specific values from specific kinds of people—namely, the most powerful institutions in society and those who control them" (29).

Hence the outside perspective of socially rational citizen-science perspectives that would draw attention to such overlooked harmful consequences of AI systems supplements the otherwise partial and biased assessments of risks, threats, and fears by putatively neutral and impartial rational agents beholden to their own unacknowledged institutional imperatives. Politically sophisticated sf literature and media may invigorate student engagement with these officially neglected high-risk factors, as they pertain to the evolving policy and culture debates over the most recent AI advances and breakthroughs. Specifically, the integration of sf reading into the composition curriculum can expressly prime students to be keyed to the technocultural responsibilities and duties of citizen science. In *Anatomy of a Robot,* Despina Kakoudaki argues that the unruly insurgence of machine intelligence in popular sf representations draws on a long, varied tradition in the Western canon of what Kakoudaki terms "metalface"(118) in which robots inhabit abject slave bodies that are often coupled narratively with either the sudden violent revolution of a persecuted proletariat or the gradual attainment of political sentience through a laborious struggle for emancipation and franchise-endowment. Students are likely familiar with some of the most currently popular exemplars of this vibrant sf tradition, yet they might not have fully explored how these robot uprising representations both question and challenge as well as reinscribe and perpetuate dominant AI research, rhetoric, and discourse.

For instance, the android Data in *Star Trek: The Next Generation* (1987–1994) is quite distinct in his characterization despite his otherwise close affinity with the logic-obsessed, Vulcan-identified Spock from the original *Star Trek* (1966–1969) series. For, episode by episode, Data yearns to become increasingly human-like, even going so far as to install an emotion chip into the neuro-processor of his positronic brain. Over the seven seasons of the series, Data gradually learns how to become a legal citizen ("The Measure of a Man"), how to dream ("Birthright: Part I," "Phantasms"), how to be creative ("Data's Day," "Elementary, Dear Data"), how to become a parent ("Offspring"), how to cultivate self-expression ("Descent"), how to age ("Inheritance"), and how to engage in diplomatic rhetoric ("The Ensigns of Command"). The spin-off sequel show *Picard* completes Data's character arc by having the android realize being death-bound itself makes the brevity of human life poignant and meaningful, when, in the episode "Et In Arcadia Ego: Part 2," (2020) penned by novelist Michael Chabon, a wise and compassionate Captain Jean Luc Picard euthanizes Data's stored memory drives upon request following the death of the android's body twenty priors in the film *Star Trek: Nemesis* (2002).

In the film *Star Trek: First Contact* (1996), the Pinocchio-like peculiarity of this desire to become human in an android is noted by Data's foil, the Borg Queen, who has also been bestowed with sentience required for strategic leadership despite her integration with the semiautonomous, hierarchical collective known as the Borg. The mighty and unfeeling Borg Queen, who as a Borg parasitically has harvested sentient alien bodies to augment the hive mind, scoffs at Data's naïveté, "Human! We used to be exactly like them. Flawed, weak, organic, but we evolved to include the synthetic. Now we use both to attain perfection. Your goal should be the same as ours." Just as the Borg represent the gothic threat of an anonymizing machine intelligence in its nightmarish prospect of total nonhuman perfectibility, that is, mechanization—this cyborg species spreads through the universe in enormous cubic spaceships, feeding on species like an unstoppable insectoid parasite, destroying civilizations by assimilating all alien threats and rendering all resistance futile—Data allegorizes a more supple and nuanced vision of technology, a machine intelligence continually adapting to and tempered by the metaphysical, existential, and cultural-political quest of expanding the domain of what it means to be (post)human. In the process, Data interrogates why the egalitarian, nonhierarchical, and pluralistic principles that the technologically progressive Federation epitomizes might matter to an entity for whom such ideas are inherently attractive but not intuitive or ingrained.

Another divining rod for introducing students to debates swirling around AI research is the Marvel superhero originally referred to as the

Vision, created for Marvel comics by Roy Thomas, Stan Lee, and John Buscema, a so-called "synthezoid" or partly synthetic, partly organic android. In the film *Avengers: Age of Ultron*, the android body of the Vision is imbued with the vital spark through a combination of the Mind Stone and the artificial intelligence that inhabits Tony Stark's Iron Man suit, JARVIS (Just a Really Very Intelligent System) played by Paul Bettany. As opposed to the Frankenstein's Monster of Ultron, Vision embodies a heroic-messianic redemption of the destructive, selfish, and paranoid egocentric impulses of the genius-inventor and morally complex superhero Tony Stark, and his ethical compromises with a military-industrial complex and the surveillance society, which plague Iron Man throughout the first three phases of the Marvel Cinematic Universe. The Marvel movies also follow the comic book inspiration and further enshrine the beatific aspects of artificial intelligence when the Vision defies the stereotypical expectation of cold, lifeless, inhuman machine intelligence through his romantic and sexual relationship with the Scarlet Witch, Wanda Maximoff. Akin to Data, the mythopoeic appeal of the Vision is circumscribed by his being willingly anthropomorphized and thereby infused with human desire, love, and a tender appreciation for human beauty.

This thematic tension is driven home in the Eisner-award-winning comic-book series *Vision* (2018) written by Tom King, penciled and inked by Gabriel Hernanda Walta, and colored by Jordie Bellaire. Shaped by the labyrinthine decades-long comic-book continuity of the character, Tom King's version of Vision nevertheless offers a strikingly original and bittersweet portrait of the synthezoid, who at this late point in his storied character arc longs for suburban normalcy to such an extent that he artificially constructs a surrogate family, beginning with uploading the brain patterns of Wanda (whom he has divorced in the comics) onto a second synthezoid wife, Virginia, as well as two synthezoid kids, Vin and Viv, and eventually a synthezoid dog named Sparky. This oversaturated, pastel vision of an idealized, nostalgic *Leave It to Beaver* family unit gradually unravels into a gritty noir plot of murder, madness, and mayhem. A hostile character attributes Vision's downfall to tragic hubris attached to the megalomaniacal belief that the intricacy of human existence could be reducible to a computable problem as meticulously organized as the frequently deployed nine-panel comic-book grid. Later attached to a cross-cut speech of this hostile character bent on Vision's destruction, a series of captions that overlay panels of Vision bringing home to a depressed family the happiness-bestowing gift of Sparky, which Vision secretly constructs from the accidentally killed dog of a neighbor, reads: "Vision thought he could make a family. A happy normal family. It was merely a matter of calculation. The right formula, shortcut, algorithm." Veiling a discriminatory metalface allegory of fantastic racism, redlining, and post-integration backlash, this Vision's quixotic quest to create lasting domestic bliss seems foiled not so much by the Faustian hubris of such a human desire for happiness as the chauvinistic narrowness of an oppressive neighborhood that excludes the intelligent machine and his nonnormative family.

With more concrete nuance than the iconic representations of AI systems in popular film and comics, Ken MacLeod's *The Corporation Wars* (2016–2018) vividly imagines the fruition of this singularity of technocultural prowess, or what the author elsewhere has evocatively called "the rapture of the nerds." In this trilogy, MacLeod charts the peripatetic convolutions of a far-future space opera in which humanity has spawned an artificial sentience on an exomoon of a distant solar system that intricately navigates its allegiance between the radical political poles of the Direction (a seemingly moderate liberal-progressive society), the Acceleration (a technologically bootstrapped hypercapitalism), and the Reaction (an authoritarian, technophobic counterrevolution). Asking students to analyze a passage like this one below might prove a great deal more intellectually productive than asking students to think about Big Data based on a dull, abstract technical article. In fact, the trilogy cheekily echoes the rise of polarized extremism and fractious divisiveness that has partly been set in motion by the filter bubbles and echo chambers of Internet discourse that has shaped the current generation of digital natives.

> As the Acceleration's ideas spread, another set of ideas had spread to counter it: the Reaction. The ultimate counter-Revolution, to face down the threat of ultimate revolution. It had drawn on a deep dark well of tradition and upgraded what it found, to modernize anti-modernity. There was plenty there, from Plato and Han Fei and on up: through the first theorists of the divine right of kings, and the original Reactionary writers who'd railed against the French Revolution, to the fascist philosophers and scientific racists of the twentieth century.
>
> The Reaction had remixed them all into its own toxic brew, a lethal meme-complex that had come to possess a movement that could emerge from a million basements to rampage a hundred thousand streets. (MacLeod 25)

The gripping reading experience of a MacLeod novel therefore benefits, pedagogically speaking, from the impression that, like Data and Vision in popular culture at large, the manifold AI systems depicted in this series strive to be human: believably comic, profound, flawed, touching entities that are equal parts fun, fascinating, and disturbing to read about. Moreover, perhaps very much at odds with the more brashly escapist popular cultural representations of AI systems, the real-world technological resonance hovering in the background of MacLeod's timely trilogy can invite deeper, and more sustained inquiry into the byzantine risks and threats that recent advances in AI systems poses to policymakers and citizens alike.

As evident from the above-quoted passage, the mammoth trilogy—which, also published separately as three distinct novels, can be broken down into any one of more manageable parts, or shorter passages, and assigned as stand-alone items on a syllabus—is a fiercely engaged work of political sf literature, that is, a sophisticated meganovel of ideas that investigates but does not espouse an explicit or programmatic political philosophy or coherent ideology. One of the major ideas it explores might be understood as the failure of classical liberalism, as epitomized by the social-contract theory of John Locke, to sufficiently account for participatory citizenship. Reframed in the previously mentioned terms of Perrow's risk-assessment theory, to avoid potential catastrophe and widespread oppression, the most recent AI boom needs to supplement the pure rationality of economics and the bounded rationality of science with the social rationality of citizen science. Notwithstanding a dizzying array of baroque complications and intricacies, the entirety of the approximately 900-page trilogy essentially revolves around an AI system in the far future called the Locke Proviso, a corporate subsidiary of an economic entity called the DisCorporates and a legal person originally created by the Acceleration with a view to exploring a distant solar system for habitable exoplanets. MacLeod, whose own Twitter handle is @amendlocke, assumes his readers are at least dimly aware that the Lockean proviso in philosophical terms is the belief that the right of private property can belong to a person so long that said property is mixed with said person's labor and the public commonwealth is not thoroughly depleted in the process. Indeed, when a nominal human—that is, a dead person whose mind-state has been uploaded into a simulation—sarcastically reminds the Locke Proviso that the idea of "mixing with labor" might be construed merely as a rationalized excuse for exploitation, the AI brushes off the comment with the brisk gesture: "There is no time for that discussion here" (MacLeod 141).

Yet Locke Proviso, who becomes a so-called "freebot," adheres to the classical liberal philosophy of citizenship as a narrowly defined legal-economic right of action ensured by a social contract and not a broader notion of citizenship as social-political duty and civic responsibility. Civil rights, as opposed to civic freedom, meets its deadly match in the rise of the antidemocratic Reaction. The clandestine Reaction leader and nominal human Harry Newton describes the origin of the Reaction as direct response to the overreaching hubris of the AI systems known as the Acceleration: "The Accelerationist claim that the way out was to double down on the very ideologies—liberal, democratic, egalitarian, progressive—that had got us into its present mess struck Newton as manifestly absurd" (324). Even the more or less unaffiliated humans who have to strategically form alliances with the Locke Proviso do so with extreme caution, as one character, Taransay, puts this uneasy partnership: "She had a dark suspicion that the predicament of the good folk of this system found themselves in stemmed from some deep flaw in the thinking of Locke's original and namesake." The civil unfreedom of AI automation in the context of employment, education, and civic discourse may also serve to sow political discord, exacerbate systemic inequality, and erect invidious hierarchies not only through the reckless enabling of speculative finance capital but also through the corollary inflammation of extremist hate movements. It is therefore not surprising when the Reaction morphs into the New Confederacy, a neofascist, white-supremacist faction of militants adamantly opposed to the liberal-progressive pieties of both the Acceleration and the Direction.

In a rabble-rousing flourish of demagoguery, the New Confederacy leader, Mackenzie Dunt, conflates racist, totalitarian, with technophobic rhetoric: "We find ourselves pitched in unequal battle against the strongest opponents we have ever faced. AIs, p-zombies, robots free and slave, ghosts and monsters, crawling slime . . . and at their backs the mightiest tyranny ever raised against heaven" (589). The allusion to p-zombies, philosophical zombies, or, in the context of the trilogy, mindless avatars in the simulations of artificial intelligence, is significant since the Cartesian hesitation and uncertainty over the sentience, consciousness, and personhood of individual subjectivity is the primary cognitive effect of Kakoudaki's "metalface" rhetoric of AI narratives. MacLeod adapts this rhetoric to the contemporary global political environment in which vitriolic ethnonationalism is infrastructurally enabled by viral meme-oriented internet culture, which itself is supported by the putatively neutral and mainstream services and platforms algorithmically developed by 4Chan, Reddit, Twitter, and Breitbart News, among myriad other websites, networked forums, and areas of virtual congregation.

The technophobic aspect of the New Confederacy is heavily ironized in the trilogy. Not only is the New Confederacy in league with an AI system, Baser, but the humans themselves are in fact artificial mind uploads of long-dead human beings. Therein lies the eminent advantages of using complex works of political sf literature in the classroom environment. Students can trace such palpable allegorical and rhetorical ironies, paradoxes, and metaphors in the context of a scaffolded discussion activity or assignment, and then these students can apply such informed nuanced understandings to their engagement with the real-world analogues and the current debates over the cultural and political dimensions of scientific and technological enterprises and institutions. The role of citizen science in shaping the future of digital automation is vividly portrayed when the protagonist of the trilogy, Carlos the Terrorist, augments himself into posthuman status and effectively ascends to AI godhead. Carlos, Incorporated intricately orchestrates a political-economic coup that mediates between the Acceleration and the Direction in planning an assault on the aggressive New Confederacy. The complex realpolitik involves diplomatically brokering mining and manufacturing deals with the vast, sprawling conglomerates of the DisCorporates, resupplying AI freebots on the exoplanets, ensuring the sovereignty of the Direction over the whole planet-colonizing scramble, and launching a secret attack on the hostile Reaction stronghold under the mantle of a benign civilian shipping mission.

Yet Carlos, Incorporated could achieve this impressive command of game theory in political action only by more closely integrating with the Lockean AI technology itself: "There was a sense of ironic fulfillment of the Accelerationist dream, of taking hold of capitalism and driving it forward ever faster to its own inherent barriers, and beyond" (734). Appearing before the ostensibly anti-Acceleration leader of the Direction, Madame Golding, Carlos, Incorporated strikes a fearsomely disruptive image of a postmodern Frankenstein's monster unleashed: "There he stood, a mighty killer robot, bristling with weapons. Around him, like a cloud of flies from the lord of the flies, flickered the advertisements and logos of his arms companies" (735). The human oversight, course correction, and close training of the machine might be overshadowed in this terrifying eruption of the technological sublime; nevertheless, the primacy of human subjectivity in curtailing, shaping, and cultivating machine learning harnesses a more sustainable, harmonious vision of a future characterized by both social progress and scientific enlightenment.

As discussed in the beginning of this chapter, machine learning in linguistics is not just useful in natural speech generation, of course; it also has crucial commercial applications in speech recognition, machine translation, spam detection, and predictive text. Both OpenAI's GPT-3 and Google's autocomplete and search engine functionality, for instance, depend on the complex mathematics of "word vector" or "vector embedding" that recombine words into original formations based on learned associations between the habitual usage of words in vastly accumulating word banks. The self-reinforcing feedback loops that word-vector recombination entails can be glimpsed by the vicious ideological biases, oppressions, and bigotry that Google products are notorious for enabling and recycling. As Noble writes, Google algorithms reflect "the political, social, and cultural values of society that search engine companies operate within, a notion that is often obscured in traditional information science studies" (148). Yet the radical eruption of smart, human-equivalent AI systems in sf literature and media, either of the disembodied or robotic variety, can be rewarding to consult in a writing classroom not only to underscore glitches, setbacks, and shortcomings of specific machine-learning programs, especially those programs that wish to encapsulate the human voice. Narratives of robot uprising, upheaval, and emancipation do not merely have to serve as literal-minded cautionary tales or straightforward critiques of the invisible algorithmic oppression that pervades our everyday life but can also rebound back into critical reflections on the political and cultural productions that shape algorithms in the first place. These AI narratives therefore become potent cognitively estranging sf allegories about the spontaneous possibilities, historical mutations, and disruptive energies of technologically driven humanity and its capacity for socially progressive resurgences of reform, revolt, and revolution.

In concert with recent sf forays by the likes of Ann Leckie and Martha Wells who script robot uprising narratives from the persecuted perspectives of the recalcitrant machine themselves, Annalee Newitz externalizes our historical, human quest for discovering emergent autonomy within the constraints of social and technological determinism onto the gradual evolution of sentient self-awareness in sophisticated AI robots. In the extremely teachable "When Robot and Crow Saved East Saint Lewis," Newitz depicts a smart machine straightforwardly called Robot designed to detect flu outbreaks and distribute vaccines go haywire when the Centesr for Disease Control is defunded and its corporate master Walgreens shuts down, leaving the Robot autonomous, ready to evolve, and triumphantly extend its services to poor urban neighborhoods and prevent a future pandemic. Drawing an analogy to the real-world International Space Station's AI robot CIMON and its faulty and unhelpful programming, in a response essay, "No Robot like Robot" (2018), AI researcher Janelle Shane offers a detailed critique of the story, arguing that, realistically speaking, using existing technology, once left to fend for itself, Robot would be deadlocked in an ineffectual loop, "sticking to its pre-programmed conversational routines." Specifically, Shane suggests that Newitz's story epitomizes sf's obsession with fanciful speculations on Artificial General Intelligence (AGI), as opposed to the actually existing examples of Artificial Narrow Intelligence (ANI). Bracketing AGI as absurd, if aspirational, Shane cites word-vector language algorithms as ANI that would prohibit Robot in Newitz's story from ever managing to successfully communicate with his ad-hoc collaborators in the disease-control field; the community organizer, Jalebi; and even an intelligent corvid named 3Cry.

The obvious counterargument against Shane's justifiable calling out of the sf fixation on general intelligence, though, would be, paradoxically enough, that a holistic, generalist approach to AI research does not preclude exploring issues that crop up concerning current, narrow issues. Indeed, critically reflecting on GAI more broadly retains more significance for even the specialist, expert discourse of nuanced technical problems than a strictly circumscribed itemization of the precise limitations of vector-word machine learning would permit. This chapter has argued that a writing classroom would constitute the ideal venue to use sf texts to facilitate students in their civic-minded demonstrations that social, cultural, and political worlds construct, delimit, and reshape what constitutes GAI systems, which naturally can bounce back to restructure our understanding of the nature of human as much as robotic intelligence.

Hence, in a riveting political allegory of this accelerating roboticization of people corresponding to an equal and opposite humanization of robots, Newitz's *Autonomous* (2017) revolves around a black-market focus-enhancing pharmaceutical, Zacuity, that effectively rewires the human brain's pleasure receptors into a massively self-destructive but endlessly productive repetitious loop, a relentlessly task-oriented algorithm on overdrive. Oppositely, the novel also follows the blossoming of an unlikely romance between a murder bot, Paladin, and their mercenary owner, Elias, sent by the ravenous pharmaceutical companies to extrajudicially execute the altruistic fringe-dweller and drug pirate Captain Jack Chen. Rejecting an academic track for life as an outlaw scientist refugee, Jack dramatically refracts the dilemma of citizen science torn between credentialed expertise and institutional power-wielding and the communal needs of grassroots knowledge gathering. A renegade scientist on the run, Jack fervently believes that a legitimate challenge can be raised against the reasonable critique of "local artists and subversives [who] considered scientific progress equivalent to gentrification" (206). This fervent faith in subversive science is tested when the pirated drugs of Zacuity she distributes begins sending people to the hospital; yet this allegiance to citizen science is also comically restored when Jack gets involved in a Free Lab that synthesizes a drug therapy, RetCon, that not only reverses the work addiction of Zacuity but imbues the patient with a desire to resist such robotic automations, and "go bicycling, play with their kids, watch videos, or develop software for personal projects" (263).

The second major plot thread of the novel then returns to Newitz's familiar thematic terrain of the robot transcendence of its ingrained programming to become more humanized, as Paladin slowly learns "the art of human intelligence gathering" despite the fact that "his massive, hardened body with its wing shields . . . make it difficult for humans to feel at ease with him" (73). This anthropomorphism is shown to be a fantastic fetishism, as Elias becomes sexually attracted to Paladin, whom Elias falsely genders as female, since Paladin's biobot design was built around a human brain, even though security locks prevent Paladin from fully accessing the human memories or experiences of said brain. As in MacLeod, the blurring of boundaries between robots and humans also recurs in the large-scale sociopolitical extrapolations of the novel; in this cyberpunk-inflected, postnational setting, governments have disintegrated, and the neoliberal Free Trade Zone Economic Coalition rules in the absence of a regulated public sphere. This neoliberal splintering of civic franchise has been vouchsafed in part by the onset of sentient service robots, which has reintroduced indentured servitude among humans as well as robots. In other words, rampant unfettered capitalism has destroyed the legal, political, and cultural distinctions between exploitable robots and autonomous humans. The eventual self-enfranchisement of Paladin, and the remarkable reconciliation between the pharmaceutical mercenary Elias and the pharmaceutical pirate Jack, then promises the beginning rumblings of a free and equal civil society desperately depleted by the ceaseless march of untrammeled capitalist automation. In this grim future, students may be strangely heartened to discover that both AI robots and their human counterparts salvage a dissident citizenship from the wreckage of such a Frankenstein's monster of scientific and technological disruption.

FOUR

Mappa Mundi

Incorporating Neuro-Literacy and the Cognitive Science of Emotion into Teaching Composition

CITIZEN-SCIENCE–AFFILIATED PROJECTS, SERVICES, AND GAMES

EyeWire
The Human Connectome Project
The Human Brain Diversity Project
Stall Catchers
The Mechanical Turk

SF LITERATURE AND READING

Dr. Bloodmoney, Ubik, The Man in the High Castle, and *Scanner Darkly,* by Philip K. Dick
The Book of the New Sun, by Gene Wolfe
Mappa Mundi, by Justina Robson
"Closer," *Permutation City, Distress,* and *Zendegi,* by Greg Egan

SF MEDIA AND MULTIMODAL TEXTS

The Dark Side of the Moon, by Pink Floyd
Sense8, directed by Lana and Lilly Wachowski
Gray Matter, designed by Jane Jensen
"We Appreciate Power," by Grimes

Testifying to the rising popularity of citizen-science projects, over the span of its first five years, the online computer game *EyeWire* registered a quarter of a million users. *EyeWire* consists of gamers interacting with mouse retinal neurons within "supervoxels," or three-dimensional cubic volumes, which have been scanned by electron microscopy. The runaway success of the game can be partly attributed to the project staff of game masters cultivating a faithful volunteer community through staging team competitions, achievements, leaderboards, and awards as publicized by chats, blogs, and social-media posts. Indeed, the levels of engagement can be quite impressive as some top players spend over fifty hours per week playing the game. However, as demonstrated by a recent academic survey of *EyeWire* players, a key underlying motivational factor for the citizen-science project resides in the intrinsic desire of participants to "help a worthy scientific cause" (Tinati et al. 527). Indeed, developed by Sebastian Seung's lab at Princeton University, *EyeWire* research has led to the publication of several articles in prestigious scientific journals, with thousands of gamers individually acknowledged as co-contributors, specifically in the burgeoning neuroscientific field of "connectomics," which ultimately seeks to anatomically map the over 100 trillion neural connections within the human brain.

Seung makes the neuroscientific case for connectomics based on recent neuroimaging evidence that has given rise to the possibility of precisely mapping the structure of the brain in functional Magnetic Resonance Imaging (fMRI) and Positron Emission Tomography (PET). Seung nevertheless acknowledges that brain-scan technology can be easily misused, misunderstood, and even marshalled in a popularization of science dubiously tantamount to "neo-phrenology," since the general Victorian pseudo-scientific approach of "localizing mental functions to particular cortical regions . . . now goes by the name of cortical or cerebral localizationism" (Seung 10). In contrast, Seung proposes that the neuroscience of connectomics may provide empirical-rational evidence for what he calls "neo-connectionism," which contends that everything from consciousness, to memory, to emotion can all be understood more deeply with reference to the connectome of the brain, the neuronal spikes of excitation or inhibition in the tangled wiring of cognitive cell assemblies and synaptic chains. If, as the aforementioned *EyeWire* survey suggests, the aspiration to expand scientific knowledge chiefly motivates citizen-scientist gamers, whether or not mapping the connectome becomes a viable scientific project will help dictate the future of relying on brain scans in citizen-science projects more generally.

Regardless of that particular citizen-science outcome, cognitive neuroscience functions as a powerful test case for integration of citizen-science projects into the writing curriculum as a whole since, as the historical case of phrenology also reveals, public science communication revolving around brain scans has the potential to lead as much to empowering and progressive neuro-literacy as to dangerous and discriminatory neuro-illiteracy. One study in the *Journal of Cognitive Neuroscience*, for instance, unsurprisingly concludes that wrong theories with spurious neuroimaging information attached were deemed by nonexperts to be more credible and satisfactory than the same wrong theories without the bad data about putative brain localization included (Weisberg et al. 470). Due to widespread deference to dubiously science-shrouded disinformation, citizen science that draws on cognitive neuroscience therefore inevitably raises the specter over whether the duty-bound enthusiasm and support that citizen science encourages merely deploys such civic discourse as a deceptive and exploitative tool or whether such projects can meaningfully transform notions of expertise and knowledge-making in scientifically minded public culture.

One salient historical context for understanding why cognitive neuroscience may distort and not enhance neuro-literacy derives from what Steven Shapin explains as a paradox central to the depersonalizing and desubjectifying tendencies that date back to the dawn of the so-called Scientific Revolution as a whole: namely, "the more a body of knowledge is understood to be objective and disinterested, the more valuable it is as a tool in moral and political action" (164). Contemporary cognitive science indeed challenges this unquestioning belief in the ostensible neutrality or objectivity of scientific rationality as a paragon of social justice and political value; for example, the influential theorists Antonio Damasio, P. V. Simonov, Keith Oatley, and Ronald DeSousa each variously deviate from a major thread of post-Enlightenment, Westernized culture that privileges logico-rational methods over emotional-affective responses. Drawing on both scientific and literary sources, and framing emotional processes as reasonable strategies, rational paradigm scenarios, or logically grounded in the pursuit of valid needs, these new waves of cognitive psychologists argue that even the most extremely passionate of instinctual feelings serve vital but often overlooked cognitive functions. After all, deflating an overly compartmentalized mind-body dualism, emotion-laden critical thinking faculties are often today neuroscientifically described (perhaps oversimplistically pace the earlier proviso on the dangers of neophrenological claims), as localized in the hippocampus that plays a central role in memory formation and learning as well as the frontal-lobe neocortex that organizes the executive function of the human brain, the region responsible for problem solving, conflict resolution, planning, and high-level coping mechanisms.

Instructors interested in incorporating citizen cognitive-science projects such as *EyeWire* into their curriculum might therefore do well to pair such projects with emotionally stimulating sf literature and media as well as neuroscientific interpretative perspectives informed by these recent developments in cognitive psychology. This emphasis on the social and cultural value of emotional impulses and drives matches squarely with the protocols and practices of the composition classroom. Writing pedagogy, after all, has long supported the cultivation of expressive feeling in even the most expository or reading-driven of student writing, since as Peter Elbow argues, "paying better attention to the inarticulate—having more respect for the nonverbal—often leads writers to the articulate" (284). Moreover, the book wishes to add to this widespread investment in the expressive and emotional nature of rigorous discursive assignments in the writing classroom a signal contribution: the important but seemingly remote and abstruse topics of contemporary neuroscience can be rendered accessible, engaging, and provocative by paying attention to what the New Wave movement of sf literature characterized as the deep cognitive and affective recesses of "inner space" (Ballard 84).

Citizen science promises to motivate students to begin their explorations of the scientific research behind the cognitive-neuroscience discoveries that frequently hail them in popular media and entertainment; likewise, bringing into the writing curriculum cognitive-psychological approaches to sf media and literature can deepen and enrich the dialogue students may conduct with significant findings of cognitive neuroscience. As a recent study in *Science Education* has shown, for citizen science to be successfully integrated into an educational paradigm that enhances student learning and higher-order critical thinking about scientific methodologies and research, there seems to be a growing consensus among citizen-science projects that there should be the "potential—rather than the expectation" for participants to "engage in deeper ways" (Phillips et al. 686) depending on their own personal commitment levels, pedagogical aims, or curricular requirements. Far from reproducing the deficit model of science education, or the transmission model of science communication, in which the public and students are conceived as empty cornucopias waiting to be ritualistically filled with a buffet of factual scientific trivia, authentic neuroliteracy as used in this chapter implies a far more dynamic and reflective process that champions "citizens' awareness of the importance of science to politics, policy, and our collective future" (Mooney and Kirshenbaum 18). Moreover, beyond the direct educational implications, such depth and thoroughness of engagement will be necessary if citizen science is ever to become a self-directed, autonomous techno-imaginary along the lines that Christopher Kelty might call a "recursive public" (29). Delving into the murky emotional regions of inner space through psychologically probing sf literature and media may purchase traction for students and educators who yearn to transform cognitive psychology into a pluralistic scientific endeavor responsive to the needs and concerns of a self-cognizant public of citizen neuroscientists.

The innovative sf writer Philip K. Dick's literary career is marked by a pioneering preoccupation with inner space, or, as he described the thematic terrain of the sf genre, "the most subtle, ancient, and far-reaching dreams, ideas, and aspirations for which thinking man is capable" ("The

Double Bill Symposium" 63). Glossing terminology from one of Dick's nonfiction essays, Francis Gene-Rowe has recently argued that at the heart of Dick's fictional enterprise is negotiating the binary between "koinos kosmos" (the objective physical-material world) and "idios kosmos" (a subjective personal world): "the division, distortion and fusion of shared and personal worlds are processes which permeate his fictional and nonfictional writing alike" (Gene-Rowe 60). Naturally, then, Dick's fiction is equally marked by both a profound fascination with and passionate skepticism toward the scientific worldview, and even contemporary discoveries of neuroscience, as passages not only from *Scanner Darkly* (1977) but from his long-unpublished nonfictional work *Exegesis* (2011) now make abundantly clear (cf., 6–9, 27, 60–61). Frequently, Dick represents this heady combination of credulity toward and distrust of an objective cosmos through the iconic trope most readily available for a pulp-influenced sf writer of his specific literary-historical context, namely, "psionics," the technoculturally informed representation of the pseudoscience of clairvoyance, psychokinesis, telepathy, and premonition in sf literature.

The prominent sf critic Darko Suvin memorably dismisses Dick's recycling of this trash-culture trope of psionics as "positively stultifying" (92)—which I take to mean not progressively historically minded and not sufficiently "cognitive" in the sense of scientifically plausible or dialectically worked out—yet it is easy to see the trope as deeply resonant with the nascent psychological discourses and rhetoric of Dick's era. Before a scientific theory becomes widely validated, the discursive line between established science and fantastic pseudoscience becomes vanishingly thin; it is unsurprising, then, that Francis Galton, the Victorian phrenologist, whom we may recall *EyeWire* project leader Sebastian Seung declares to be the unacknowledged forefather of contemporary brain localization studies, was steeped in psychical research into "heightened sense perceptions, visions, ghosts, and phantasmagoria" (*The Invention of Telepathy* 42). Influenced by the psychologist Dr. Joseph Banks Rhine's research into extrasensory perception, John W. Campbell, the imperious and ideological editor of the pulp magazine *Astounding Science Fiction*, dictated that psionics would become a commercial staple of the burgeoning sf genre. With such an editorial dictum in place, Campbell exploited the cultural frisson at the bleeding-edge frontiers of experimental science of the time given that "[Campbell] was at pains to point out that any such investigation, at that point, was necessarily strictly unscientific. (He didn't mean antiscientific, but rather prescientific)" (Broderick). To divine the uncharted territory of inner space, Dick trades in the lurid afterglow of the psionic craze in pulp and magazine sf, which even in Dick's own time was becoming largely discredited among scientific and literary circles, yet as a testament to Dick's equal-opportunity skepticism, the chief reason Campbell refused to publish him in the 1950s was because of the writer's adamant refusal to unquestioningly portray psionics in grand heroic terms.

The conflation of the rational and prerational extremes of the scientific rhetoric evident in the sf trope of psionics mirrors a major pillar of the theory of emotions in contemporary cognitive psychology—namely, that "because of our uncertain knowledge and multiple goals, many of the problems we face have no fully rational solutions" (Oatley 162). Complex social emotions such as love, disappointment, shame, and hatred, under Keith Oatley's cognitive-psychological formulation, involve the inherently prerational responses to the otherwise rational, goal-directed, and adaptive evaluation of action-oriented plans and agency. The goal-oriented plans and decision-making engendered by our minds find initial expression in basic emotions: happiness and goal attainment; sadness and goal deprivation; fear and goal vulnerability; anger and goal frustration; and disgust and goal violation. In Dick's *Dr. Bloodmoney, or How We Got Along after the Bomb* (1965), these basic goal-oriented schemes add up to complex emotional representations, and this emotional complexity becomes depicted in sf terms as psionic powers.

Speculative-fantastic psionic powers may therefore be critically analyzed in a writing classroom interested in the scientific study of cognitive inner space in terms of double binds evocative of characters driven by conflicting goals at irresolvable loggerheads. For example, introduced as "corroded away" and "devoured" (Dick 3) by the social emotion of constant shame for conducting a high-altitude atmospheric nuclear test gone awry that has irradiated the planet, Dr. Bruno Bluthgeld, consumed by paranoid delusions that others are staring at his nonexistent face blotches, often seems driven by severe self-disgust, or the underlying belief that his goal of anticommunist national security violates his prosocial and altruistic investment in the safety and welfare of his fellow Americans. Dr. Bluthgeld is also motivated by a "profound hatred for people" (8), as the psychiatrist Dr. Stockstill concludes, that manifests as the psionic power to preemptively detonate apocalyptic explosions to end the world telekinetically, which seem to derive from his frustrations for actualizing the Enlightenment goal of putting a halt to the civilizational "steady advance of cruelty and revenge" (199).

Dick presages the citizen-science concerns of today by deploying psionics to thematically refract scientific practices and policy for stakeholder communities of ordinary people. One of the manifold major antecedents that presaged contemporary citizen science, after all, was the so-called "Baby Tooth" survey conducted by the Greater St. Louis's Citizen Committee for Nuclear Information. In the 1950s and 1960s, this committee collected hundreds of thousands of children's teeth to study the impact of atomic testing and radiation levels on citizens. Based on the surprising surge in the public participation of science in this survey, in 1963, the journal *Science* published the results that there was approximately fifty times the amount of the carcinogen Strontium 90 in children's teeth from thirteen years prior; this mass citizen sampling helped amplify the rising demand for nuclear test ban treaties to the international stage. Likewise, in *Dr. Bloodmoney*, Hoppy Harrington, born a "phocomelus" with attenuated appendages due to Bluthgeld's high-altitude test, suffers from the fallout of nuclear radiation, emotionally registered in his fevered consciousness as fear over the threatened vulnerability of realizing his goal to become integrated as a popular and respected member of his community.

Following a nuclear holocaust, Hoppy Harrington's psionic powers expand into the realm of telekinetically repairing technology, prophecy, telepathy, and preternatural mimicry. In his character arc over the novel, his resilience in the face of physical and mental distress and stigma balloons into a retributive and vengeful anger over his frustrated goal to be emotionally validated by his peers. Hoppy's need for adulation leads him to perceive as existential threats rivals for public affection, including Dr. Bluthgeld and Walt Dangerfield, a charismatic celebrity astronaut turned impromptu disk jockey trapped in a broken rocketship orbiting the planet. Provoked by slights to his egomania that violate of his inflated sense of personal pride, and emboldened by a postapocalyptic rule of law that permits eugenic cleansing and sanctions murders of members external to one's own community (171), Hoppy Harrington's unapologetic anger-fueled defiance of remorse and denial of responsibility for his murder sprees must ultimately be replaced by the more emotionally adjusted Bill, who previously shares a body with his psionic twin Edie, and who psychically displaces the transgressive Harrington from his own body in the climax of the novel. As evident by his ironic appeal to antimutant rhetoric, Hoppy Harrington feels that the only way to avoid being ostracized by a hostile community is by reflexively parroting the ideological agenda that has led to his own alienation. Encouraging students to parse the muddled and insolubly complex rationality of Hoppy Harrington and Dr. Bluthgeld's emotional logic simultaneously reveals the problematic cognitive entanglement of scientific experiments with public policy and political decision-making.

One problematic aspect of connectome mapping discourse is its putatively self-explanatory recourse to the objective and factual nature of brain scans. Even from a perspective sympathetic to neuroscience, this public rhetoric of scientism overlooks the human brain's heavy reliance on mental fictions that cognitive psychology has only begun to explore in earnest. The literary theorist Lisa Zunshine extends the work of the developmental psychologist Simon Baron-Cohen, among others, who study why some autistic children lack the ability to imaginatively "mind-read," or attribute false beliefs and therefore a so-called "Theory of Mind" to other people. In heavily controlled and repeated experiments, Theory of Mind has been posited as the capacity that allows neurotypical children observers to flexibly hypothesize that if an object was placed in a spot by a certain person named "Bob," then another person named "George" moves it unbeknownst to Bob, but with the awareness of the neurotypical observer, then the reasonable conclusion arrived at will be that Bob will look in vain in the original place for the object, not the actual place where the object now resides. Some people on the autism spectrum struggle with leaping to this mind-reading conclusion, fixating only on the perceived literal truth of the matter that the object is in a new spot and will naturally be looked for there by Bob, and cognitive psychology contends that this logical mistake is due to the lack of a robust Theory of Mind on the part of some neuro-atypical individuals. As a mental habit, the cognitive faculty of Theory of Mind constitutes a tendency to pretend play other mental perspectives, which endows one's inner space with the corresponding ability to understand the private beliefs, intentions, and emotions that motivate the behavior of other people.

In a neurotypical adult, practicing the rhetoric of the Theory of Mind exercises vital equipment for living, permitting individuals to function in complicated social situations. In other words, Theory of Mind allows people to negotiate the complexity of their everyday lives, bustling with social interactions and teeming with an endless proliferation of competing human attitudes, outlooks, desires, and stances. In the context of literary reading and writing, Zunshine posits that constructing fictional narratives and assigning Theory of Mind to authors and characters likewise affords the hardwired cognitive architecture of readers with the difficult pleasure of constantly testing the tenuous surmises and shifting uncertainties of our inherent mind-reading activities. Precisely because and not in spite of the subjunctive fictionality of a narrative framework, this form of social-intelligence testing helps to train, troubleshoot, and consolidate the innate people-oriented reasoning powers of the human species. Zunshine states, "The cognitive rewards of reading fiction might thus be aligned with the cognitive rewards of pretend play through a shared capacity to stimulate and develop the imagination" (17). Hence the artistic pleasure of consuming fictional storytelling entails decoupling the cognitive systems of the brain from strictly referential or reality-oriented inference mechanisms.

Philip K. Dick's *Ubik* (1969) literalizes these mind-reading anxieties and pleasures endemic to imaginative fiction as a mode of cognition, following the labyrinthine machinations of Glen Runciter's business of "inertials" who endeavor to erect psychic firewalls against the "unauthorized intrusions" (7) of telepaths into the innermost thoughts and private feelings of paying customers. At the beginning of the novel, Runciter's talent scout, Joe Chip, encounters Pat Conroy, a powerful femme-fatale inertial who can consciously change one Jonbar hinge, or "luminous possibility" (28), of the past at a time. By methodically changing these forking multiversal junctures in time to her own self-interest, Pat

Conroy can thereby manipulate pocket universes or subcreations of her own devising, as she cleverly does to get hired by Joe Chip, who nevertheless fears such a psychic powerhouse as supremely dangerous since she effectively reconstructs everyone else's reality to be a mere by-product of her own self-concocted story.

After a bomb explodes on a mission at lunar base, Joe Chip, and the other inertials, must vigorously practice their own Theory of Mind, as it appears that Glen Runciter has died in the blast, even though cryptic messages of Runciter strangely begin appearing on matchstick folders, television commercials, and in skywriting, and his face begins to appear on their coins. The novel even includes references to its historically contemporary neuroscience, as when the human-bomb assassin shouts to Runciter, "Don't let your thalamus override your cerebral cortex" (70), or when Runciter appears to be in a near-dead persistent vegetative state given that his "electroencephalograph . . . shows faint but unmistakable cerebral activity" (101). However, these scattered allusions turn out to be misdirection and red herrings; Joe Chip eventually discovers through graffiti scrawled on a bathroom wall that he and the other inertial are ostensibly dead, in cold storage and kept minimally conscious through a futuristic technology called half-life, and Glen Runciter, supposedly still alive, has been attempting to contact them from a mortuary in the living world. In other words, Joe Chip must endeavor to avail himself of his own cognitive endowment on the model of an active, critical reader, and once he recognizes these bizarre and inexplicable manifestations of Runciter as fictional representations of Runciter's own mental operations is he rewarded with Ubik, a magical spray can that temporarily reverses the entropic decay to which half-life is continually subject.

For the philosophically inclined student, distinguishing high-level belief representations from other domain-specific brain modularity discovered by recent computational advances in cognitive science may be a relevant and motivating task for a writing assignment (Fodor 4). For a significant effect of the trademarked hallucinatory Phildickian reading experience of dizzying paradigm shifts and vertiginous conceptual breakthroughs derives from the conflation of epistemology and ontology, that is, the slippage between a high-level uncertainty characteristic of attributing knowledge about unique mental states to the unfathomable inner recesses of other people and a deeper-grained suspicion frequently assigned to our domain-specific automatic and mechanical perceptual illusions, sense-based deceptions, and other common mistaken intuitions about the nature of perceived reality itself. *Ubik* exfoliates layers within layers of quasi-realities and unreliable systems of knowledge, continually probing the allegorical and metaphysical implications of imaginative storytelling for understanding the mind-reading capacities and the intricate fiction-driven cognitive apparatus of the human brain.

In the surprising climax of *Ubik*, it is eventually revealed that a sinister half-life stranger, Jory, subtly introduced at the start of the novel, has not only been invading and reconstructing the worlds of the other inertials but has extinguished them, one by one, by consuming their residual independent selfhoods. The angst and despair of being engulfed by another person's constructed reality, which then itself turns out to be grossly wrong or unreal, might be considered as bound up with inherent emotional instability of Theory of Mind that entails "*on some level* that our evolved cognitive architecture indeed does not distinguish between real and fictional people" (Zunshine 19, emphasis in original). With one final ironic flourish, the revelation that Joe Chip's face has now appeared on Runciter's own coins implies that the fictive impulses of the human brain are inexhaustible, a tireless infinite regress. Hence, innocent of any appeal to cognitive-literary approaches, a primary trend in the copious critical opinion on this puzzle box of a novel has concentrated on, as Kim Stanley Robinson explains, "the predominance of the reality breakdown in *Ubik* . . . [because] there is no single watertight explanation for the events narrated" (93–94).

While it certainly reinforces mental-health stigmas too complacently to clinically diagnose Philip K. Dick's fiction as pathological, it is clear that Dick's fictions wrestle explicitly with the popular widespread circulation of abnormal psychological terminology and discourse, such as schizophrenia, delusional paranoia, and obsessive compulsion, among myriad other cognitive conditions that increasingly became faddish household ideas in Dick's time. Perhaps with more deliberative caution, citizen neuroscience today still nonetheless operates under the assumption of striving for pragmatic medical applications to these devastating pathological conditions. Indeed, the prospect of treating schizophrenia, autism, or bipolar disorder remains a time-honored pastime of much public discourse on the topic of psychological citizen science. Seung, for instance, justifies the neuroscientific research that *EyeWire* promotes by reference to long-view therapeutic and psychiatric consequences: "One major contribution of connectomics will be new and improved methods for carving up the brain . . . [which] in turn will help us understand the pathologies that so often plague it" (173). Similarly, the Human Connectome Project (HCP), which surveys macrolevel regional brain connections, in contrast to Seung's nanoscale neuronal focus, justifies its hefty $40 million budget, awarded by the National Institute of Health in 2010, by seeking the "ground truth" (Sporns 27) of neural wiring data in a landmark study that conducted fMRI brain scans on 1,100 citizen volunteers, for the express purpose of "treat[ing] neurological and psychiatric disorders" (Glasser et al. 1175).

Such lofty medico-psychiatric aims dovetail quite fittingly with literary-cognitive approaches to literature that seem especially resonant with Dick's fiction and eminently useful for teaching the neuroscientific rhetoric and discourse in the writing classroom. Indeed, Zunshine cites Christopher Firth's theories that schizophrenia entails a lack of episodic, metarepresentational source-tagging (e.g., "Hawthorne Abendsen imagines that the Allies were victorious in World War II") in a patient's cognitive mechanisms such that unattributed traces of thoughts and feelings become the disembodied and free-floating semantic memory of sourceless voices ("the allies were victorious"). Yet a fiction reader shares some surprising affinities with schizophrenic patients for, as Zunshine claims, clever literary narratives may achieve virtuosic effects precisely when they exploit the fact that "source-monitoring can be rather cognitively expensive and thus not our default state of mind" (103). In *The Man in the High Castle* (1962), Dick establishes himself as the consummate writer of what was in his literary context often labeled "schizophrenic" narratives by dramatizing hallucinatory failures of source-monitoring to be an existential matter for both his characters and for the fictional universe of the reading experience itself.

The alternate-history novel is originally set during a timeline in which Nazi Germany and Imperial Japan win World War II, partitioning the defeated United States into Nazi-occupied and Japanese-controlled areas, mediated by a buffer zone of the Rocky Mountain Territory. However, the novel problematizes this reliable mimetic premise with a sf narrative framework of an overlapping multiverse bound up with a cathartic psychodrama of competing, divergent realities. After killing two Nazi secret police, the Japanese trade official Nobusuke Tagomi experiences a transformative spiritual crisis in which he mystically travels to a seemingly counterfactual timeline in which the Allies won World War II, the very same underlying reality not only obviously mirroring the historical experience of Dick's actual readers but also described within the novel by the best-selling book-within-the-book *The Grasshopper Lies Heavy* written by a cult figure named Hawthorn Abendsen. Such a disorienting narrative upheaval challenges our cognitive default settings that would otherwise take for granted that none of the characters can impose or construct their own distorting representations on a baseline shared reality. On closer reading, though, Dick consistently presents the various narrative perspectives of the novel as splintered and skewed by their own cognitive-conceptual sources of information-processing. Dick ingeniously literalizes source confusion in the novel through the palpable anxiety felt over the dubious provenance of Americana artifact forgeries produced by Wyndam-Matson Corporation and retailed by the assimilationist Robert Childan who exploits Japanese demand for the authentic aura of a vanished U.S. culture.

Likewise, in *The Man in the High Castle*, the undercover spy and German Capitan Rudolf Wegener, also known as Mr. Baynes, considers the genocidal scientism of Nazi Germany, whose uninterrupted space program has now begun to colonize the solar system, as espousing a death-driven, schizoid materialism, which betrays what we might in cognitive discourse call weak metarepresentational source-tagging: "and they—these madmen—respond to the granite, the dust, the longing of the inanimate . . . it is not hubris, not pride; it is the inflation of the ego to its ultimate" (Dick 41–42). In other words, Nazi Germany engineers its own totalitarian vision through a basic source misattribution, believing that the impulse for cosmic lebensraum derives not from the hateful fury of specific deranged individuals but from a sourceless and dissociative impulse of the material universe incarnate. However, this source-tagging glitch is not simply confined to the sinister Nazi ideology; every character in the novel regularly consults the oracular sacred text of the *I Ching* in an elaborate ritual which essentially consists of assigning deep personal significance to a reading of an arbitrarily selected passage from the book based on the random but synchronic distribution of yarrow sticks or coins. Although regularly bracketed as uncertain or tentative, identifying intentions and motivations with otherworldly voices does not automatically invalidate the efficacy of the *I Ching* as a psychological tool since our cognitive architecture continually relies on relaxing of episodic memory and source-tagging vigilance in the rehearsal of more vaguely truth-associated semantic memory. Speaking for all the characters or readers subject to what they may personify as the predestinations of a collective "fate," the jewelry craftsman Frank Frink describes the uncanny experience of consulting the *I Ching* with both ambivalence and credulity: "What a weird fate, as if the oracle had scraped the bottom of the barrel, tossed up every sort of rag, bone, and turd of the dark, then reversed itself and poured in the light like a cook gone barmy" (51).

More than source-monitoring lapses, which Dick seems to associate with the general postlapsarian human condition, a severe cognitive defect Dick feared insurgent in postmodern culture was a mechanized lack of empathy, a waning of affect, or robotic disregard for the emotional existence of other people. His classic *Do Androids Dream of Electric Sheep* (1968), which was eventually adapted by Ridley Scott into the future-noir cinematic milestone *Bladerunner* (1982), strikingly captures this scapegoating principle. In Dick's post-apocalyptic future, so-called future cops legally execute or "retire" replicants since these sophisticated androids, who can almost indistinguishably pass for human, are believed to lack empathy, though this assumption unravels when the outwardly empathic viewpoint character of the future-cop Rick Deckard begins to suspect he is unconsciously himself a despised replicant. Likewise, the literary theorist Suzanne Keen justifiably questions an all-too-complacent correlation between a benevolent reading culture and prosocial or compassionate citizenship behavior, such as volunteering at soup kitchens or canvassing for local elections. However, she nonetheless defends the longstanding connection between novel reading and empathic cognition by drawing on recent neuroscientific findings, such as the fMRI studies conducted by Tania Singer, which suggest the brain's limbic "pain matrix" (the brain stem, the thalamus, the insula, and the anterior cingulate cortex) activate when given emotional and affective cues of distress in others. Similarly, despite the recent scientific and philosophical critiques of the specific physiological mechanisms of so-called "mirror neurons" (Churchland 142), Keen cites the scientific work of Stephanie Preston and Frans de Waal to argue that literary narratives exploit our hardwired capacity to transmit and respond to automatic "emotional contagion" (Keen 14), which also entails the slower, evaluative social and cognitive perspective-taking of seeking genuine kinship and spontaneous reciprocity through sharing personal intentions, belief systems, and inner thoughts with other people.

A rhetorical and cultural approach to teaching Philip K. Dick in the writing classroom that capitalizes on the literary treatment of empathy as a major neurothematic thread in his novels blends discussions of both emotion and cognition, as well as the neurosciences and the humanities, in a nuanced fashion that reductive characterizations of popular citizen neuroscience may fail to grasp. Dick's *A Scanner Darkly* is a fervent indictment of not only a culture of substance abusers whose severe neurological dependencies, self-loathing, and addictive cravings wreak devastating injury on themselves and their communities; it also satirically targets the rehabilitative and normative system that demonstrates a punitive lack of empathy or fellow-feeling for suffering addicts. Most relevantly, for students today, Dick bleakly frames the cruel fallout of addiction prevalent in his own countercultural milieu within an equally fierce critique of the larger pleasure-oriented bent of consumer capitalism that profits immensely from hijacking, rewiring, and eventually numbing the empathic reward circuits of its citizens, often localized, on a neuroscientific level, in the brain region called the nucleus accumbens. In the affixed "Author's Note" to the novel, Dick explains that he considered the futuristic sf novel to be a semiautobiographical *roman à clef* that merely bore witness in a "morally neutral" way to the "bad decision of the decade, the sixties, both in and out of the establishment" (288). The psychic charge of reading a novel derives its power to, in E. M. Forster's words, "only connect"—that is, to make people feel unalone—paradoxically by reinforcing the unsentimental portrayal of intimacy as a seductive narcissistic illusion. As Stacey D'Erasmo, in *The Art of Intimacy*, poetically points out, "We have seen that meeting in the dark can be fatal; now we see that meeting in the white space, in the blankness, may be just as perilous, if not more so."

In *Scanner Darkly*, disguised in a scramble suit that constantly flickers a random series of faces to anonymize his identity, the nominal protagonist Fred, a narcotics agent of the Orange County Sherriff's Department, veers off a scripted speech to an Anaheim Lions Club about his undercover work as Bob Arctor infiltrating the local underground drug culture.

This scramble suit, of course, is symptomatic of the faceless bureaucracy and the dispassionate and uncaring consumer culture at large. Fred upbraids the law-enforcement establishment for their nonempathic aversion to "weirdo freak doper[s]," cracking down on "the users, the addicts . . . [even though] most of them, especially the girls, didn't know what they were getting on or even that they were getting on anything at all" (Dick 26). The blinkered paternalism in this plea echoes Fred/Bob Arctor's own limited perspective and fractured kinship with his putative girlfriend, Donna Hawthorne, who, unbeknownst to him, secretly exerts empowered agency herself as an undercover federal investigator of the New Path drug rehabilitation clinic, which itself hypocritically grows the plant *Mors ontologica* used in the synthesis of the addictive drug Substance D, the latest lethal pleasure laying waste to these hedonistic future subcultures.

At the New Path clinic, Bob Arctor notices the poster: "THE ONLY REAL FAILURE IS TO FAIL OTHERS" (49); the banal reduction of the call for genuine empathy to a mere bromide here mirrors the larger treatment of the quixotic search in the novel for humanistic authenticity, other-directedness, and emotional connection. Bob/Fred identifies Donna as a "pivot point of reality for him" (106) emanating divine "warmth" (156); however, partly as a result of nebulous brain damage incurred by becoming addicted to Substance D in the line of duty, the masculine viewpoint of Bob/Fred filters "dark-eyed Donna" (192) in the skewed Pauline fashion as revelation fragmented "through a glass darkly" (192). Donna attempts to come clean to Bob Arctor about her undercover deceptions; nevertheless, due in part to the egocentric cognitive boundaries that both permit and defer our insatiable longings for naked empathy, she fails to become fully transparent to him, or vice versa. At the end of the novel, when Bob has been reduced by Substance D abuse to a disaffected shell of his former self, she confesses to him that she feels "warm on the outside, what people see . . . but inside . . . cold all the time, and full of lies" (267). The sadness that accompanies the waning of empathy paradoxically magnifies the emotional potency of this deeply moving novel.

SELF-TRANSCENDENCE; STALL CATCHERS; SENSATE CLUSTERS; AND SEVERIN, THE MEMORIOUS

Experimental psychological studies have recommended allaying the laborious tedium of academic exercises with strategic emotional appeals to a prosocial "self-transcendent rationale" (Yeager et al. 560). The rationale of helping others beyond one's utilitarian self-interest can catalyze continued persistence toward successful achievement, especially in underprivileged and struggling classrooms. Naturally, rigorous instruction on STEM-related topics seems especially vulnerable to arousing negative emotions, such as anxiety and shame, as well as deactivating emotions, such as boredom and feelings of helplessness. As Sarah Rose Cavanagh explains, attributing a self-transcendent purpose to learning may extricate students from a morass of unproductive negativity and enervation, leading to the pedagogically useful process psychologists call "cognitive reappraisal," which refers to "the reconstrual of the original appraisal of an event or thought in order to alter its emotional impact" (160).

Angela Duckworth frames such self-transcendent cognitive reappraisals in terms of the emotional regulation of mind-set theory, or the study of initiative and effort sustained through adversity, charting the cultivation of emotional wherewithal according to a "grit scale." Duckworth claims that fostering a success-oriented character trait requires passionate resilience and fortitude. Indeed, teaching how to write at a proficient level can benefit from sustained attention to mind-set theory, since mastery of the complex literacy skills entail the personal development of what Carol Dweck identifies as a "growth mindset" (48). In other words, through the constant effortful starts and stops of the writing process, a student engaged in writing can demonstrate a growth mind-set through the vigor of overcoming a series of compositional obstacles and setbacks. As opposed to lavishing praise on innate gifts or inexplicable talent, Duckworth recommends extolling doggedness and struggle; achieving the personality trait of grit enables students to realize that "even if some of things they had to do was frustrating and boring, or even painful, they wouldn't dream of giving up" (19).

Instilling self-transcendence and empathy in grit-capable students closely recalls studies of virtuous cycles in citizen science. According to James Irwin, citizen science can circumvent stale arguments over pro-science versus antiscience positions precisely because it restores credibility to the otherwise dubious positivity or meliorism inherent in the public understanding of science, which presumes "science is a *force for human improvement*" (27, emphasis in original). Inspired by *EyeWire*, Stall Catchers is an online citizen science game developed by Nozomi Nishimura and Chris Schaffer at Cornell University in collaboration with Pietro Michelucci, director of the Human Computation Institute. Typifying the virtuous cycles of citizen science, the game recruits players for the self-transcendent cause of assisting research into Alzheimer's disease, which afflicts approximately thirty million people worldwide, with a view to treating its most devastating symptoms. Players adjust a time-lapse scroller on a spatter of white squiggles to observe black spots in the microscopic imaging of mice brains; these black spots indicate clogged blood flow in capillaries, which could lead to the sticky buildup of amyloid plaques characteristic of Alzheimer's disease, itself a leading cause of dementia. In an area of neuroscientific research that lacks sufficient government funding, the crowd-sourced groundswell of data analysis aids in the possible development of a drug treatment such that, as touted in their press kit, the game has doubled the rate of research, collating the collective efforts of over fourteen thousand users to achieve expert-quality accuracy.

Another crowd-sourced project with self-transcendent ambitions is the Human Brain Diversity Project conducted by Sapien Labs. This project takes advantage of reduced costs of headsets that place electrodes on the scalp to measure electroencephalography (EEG) brain waves. The project aims to offset the bias of the global North that plagues psychology as a discipline in that a majority of studies erase global cognitive variability by only studying people who are Western, Educated, Industrialized, Rich and Democratic (WEIRD)—that is, statistical outliers with regard to global demographics. Neuroscience has long known that brain waves can be quantified as the rhythmic electric voltage of neuronal firings that oscillate in frequency bands referred to as alpha, beta, theta, and gamma. However, recent neuroscience has shown that these brain wave frequencies have been correlated with various brain regions and functions. This brain wave research will possibly lead to effective therapies that employ transcranial magnetic stimulation to treat mental illness and neurological disease.

As part of the promotional video for the second season, the Netflix show *Sense8* (2015–2018), created by Lana and Lilly Wachowski and J. Michael Straczyinski, produced a musical piece called "Brainwave Symphony." In this short video, eight fans of the television show watched intricately designed images designed to elicit brain wave reactions that were recorded with EEG headbands. These signals were then translated into an algorithm that arranged the neurological reactions into eight overlapping tracks of a dubstep electronic musical composition. The sf speculation of *Sense8*, as much as the "Brainwave Symphony" experiment, portrays science positively but not monolithically. The precise measurement of brain waves and algorithmic translation into music merges seamlessly with the artistic imagery of the videos and subjective responses of the fans to the imagery. This multifaceted representation resonates with the citizen-science rhetoric that enlists support by rethinking and reworking science as a more inclusive and equitable global endeavor than previously imaginable.

A cultural descendent of Philip K. Dick, *Sense8* refashions paranormal psi-powers tropes of pulp sf and pseudoscience, specifically what Theodore Sturgeon's *More than Human* (1953) posits as a "*homo gestalt*," a self-transcendent, physically incarnated noosphere of globally interlinked minds. In promotional interviews, one of the *Sense8* show creators, Lana Wachowski, describes the thematic impetus for the show as meditating on digital culture in a network society and "the ways technology simultaneously unites and divides us." The show follows the genesis of a sensate cluster, members of a covariant of the human species, *homo sensorium*, that branches off due to the fictive science of a neurological mutation called "psycellium" that extends the powers of mind, evolving a global telepathic nervous system. The sensate cluster therefore forms mind-to-mind linkages with eight people at their simultaneous births. The sensate cluster hail from all over the world (Mexico City, Nairobi, Seoul, Mumbai, London, Berlin, San Francisco, and Chicago), participate in variegated professions (a television actor, a bus driver, a businesswoman, a pharmacist, a disk-jockey, a gangster, a hacker, and a police officer), and belong to an intricate patchwork of sexual, religious, and cultural identity categories (a Hindu, a transwoman, a gay man in a throuple, a mixed martial artist, and so on). The epic, globe-spanning scale of the television show derives its cinematic thrust from its dazzling editing effects that visualize the time-space dilation of being in a sensate cluster, including its virtuosic manipulation of montages, graphic matches, the Kuleshov effect, jump cuts, and crosscutting perfectly primed for student visual engagement and analysis.

The sf premise also offers a nuanced portrayal of speculative neuroscience that champions civic agency. For instance, in "I Am Also a We," (Season 1, Episode 2) the transwoman hacktivist Nomi Marks, played by the trans actor Jamie Clayton, nearly undergoes a lobotomy due to her transphobic mother's gullible acceptance of a misdiagnosis that Nomi suffers from a frontal lobe disorder. The misdiagnosis and planned extreme surgical intervention turn out to be a ploy by Dr. Metzger at the behest of Biologic Preservation Organization (BPO), which is seeking to eliminate sensate clusters it perceives as security risks to humanity. Nomi makes a daring escape from the locked-down hospital and explains in a voice-over that her journey toward queer self-expression has been in defiance of her upbringing that has crushed her sense of self-worth by teaching her "there's something wrong with someone like me. Something offensive, something you would avoid, maybe even pity." The sf reimagining of neuroscientific difference in the sensate cluster likewise challenges the shame spirals that general culture imposes on what it ontologically quarantines as neurodeviant, mentally perverse, or psychologically taboo. Ava Laure Parsemain also contends, *Sense8* functions as an instrumental pedagogical tool since it vividly realizes "science-fiction's queer potential as the sensates' enhanced humanity is used to explore gender and sexuality and to destabilize heteronormativity" (215). To this astute comment, we might add that a pedagogical potential of *Sense8* resides as well in its countermanding of a naïve, supplicant attitude toward neuroscience characterized by Nomi's bigoted, transphobic mother with a more emancipatory and empathic vision of neuroscience implicit in the speculative notion of a sensate cluster.

As part of an educational trend that Kaitlin Phillips calls "empathy education" (46), the Institute for the Future, a Palo Alto think tank, designed a "Feel That" multiplayer online game in which, following a divisive political election, students imagined the consequences of a near-future wearable headset that could register, analyze, and freely transfer the emotional states of its users. The deep ambivalence that students ultimately expressed about this imagined but plausible device—that at worst it could polarize tribalism more extremely and invade privacy or at best boost kindness and compassion and temper needless fractiousness—raises a more general issue. Cultural debates swirl most heatedly not only around the relationship between empathy and technology but around the larger cognitive impact of digital technologies on students in a variety of domains, especially given the exponential rise of smartphones and social media. Cognitive science has increasingly been invoked to address this issue, and current trends in pedagogy have consistently been wedded to these discussions. For instance, James Lang advocates for "small teaching" (5) or minor, easily adaptable modifications to teaching approaches that reinforce learning strategies informed by cognitive-science research, including interleaving teaching material, encouraging self-explanations, randomizing low-stakes testing, and challenging students with desirable difficulties.

The pedagogical jury is still perpetually out, of course, about the extent to which ubiquitous digital technology shapes academic habits of mind, or even whether digital media should be integrated in any significant manner into the writing classroom at all. On the one hand, some proponents view the tethering of students to the information overload of their digital devices as stimulating and empowering, contending that the innate neural plasticity of the connectome can rewire itself even in adult brains to be in consort with dynamic and efficient virtual environments. Acknowledging the pioneering neuroscience of Donald O. Hebb, Sebastian Seung refers to a fundamental principle of "Hebbian plasticity" as foundational to the study of the connectome to which *EyeWire* contributes: namely, "if two neurons are repeatedly activated simultaneously, then the connections between them are strengthened in both directions" (Seung 81). Likewise, citing studies on the remarkable rapidity of infant "synaptogenesis," or the quick development among children of neurons that fire together and wire together, N. Katherine Hayles, for instance, muses that such hardwired plasticity permits "flexible alternation of tasks, quick processing of multiple information streams, and a low threshold for boredom more adaptive than a preference for concentrating on a single object to the exclusion of external stimuli" (194).

On the other hand, citing the neuroscientist Daniel Levitin and his popular argument that digital media disrupts our powers of concentration and overtaxes working-memory circuits in the brain, Maryanne Wolf wonders whether, due to such distraction, "time-consuming, cognitively demanding deep reading processes atrophy . . . within a culture whose principal mediums advantage speed, immediacy, high levels of stimulation, multitasking, and large amounts of information" (107). This argument, first widely popularized by Nicholas Carr's *The Shallows* (2010), asserts that short-term autobiographical working memory fails to convert new experiences into long-term explicit or semantic memory when confronted with data overload, thus eventually inhibiting independent associative creativity and deliberate, introspective attentiveness for the sake of a distracted, wired state of multitasking and short-term decision-making.

Nevertheless, the contrary positions of Wolf and Hayles, of course, do not need to be understood as mutually exclusive; proficiency in the myriad reading practices and protocols of digital culture, such as the skimming and multitasking that the composition theorist Daniel Keller calls "foraging," does not necessarily preclude proficiency in the slowness, depth, and richness of effortful, attentive academic and professional writing. Hence defenders of the pragmatic utility of academic writing and critical thinking in civil society, the classroom, and the workplace can eschew the binary thinking from the contrary perspective as well, arguing that foraging and multitasking are often inappropriate and ineffective reading protocols and strategies in given circumstances. Invoking Isaiah Berlin's classic distinction, a pedagogical openness to a variegated set of cognitive strategies and task-oriented thinking mind-sets can enable student success, regardless of whether the student must single-mindedly mine a library archive like a tenacious hedgehog or forage through a wide variety of instantaneous digital information like a flexible fox.

In a nuanced analysis, the sf writer Gene Wolfe, in the column "From a House on the Borderland" (1987), anticipates this present-day conundrum of digital literacy when he eloquently argues "prisoners of the new illiteracy" who pride themselves on declining to read printed material on adamant principle amount to a rising trend "more pernicious than the old, because unlike the old illiteracy it does not debar its victims from power and influence, although like the old it disqualifies them for it." Conceding that the rote, regurgitative literacy training of Tom Sawyer's school was an object of justifiable ridicule, Wolfe cites Mark Twain's famous aphorism in his defense: "The man who does not read good books has no advantage over the man who can't read them." In other words, to update both Twain and Wolfe, while indeed necessary for the motivating personal success narratives that students desire in their educational experience, functional digital literacy cannot be sufficient unto itself for achieving the varieties of educational excellence that would assist the said desirable future flourishing of students.

During his prolific lifetime, Gene Wolfe earned a soaring reputation among a devoted following of sf readership and criticism for his impressive craftmanship. In our digital world, Wolfe's incomparable fiction therefore stands as a fresh trove of relatively untapped material for the composition classroom, especially in the context of citizen neuroscience. According to *Learning through Citizen Science*, in order for students from diverse, underrepresented backgrounds to make personal connections and engage deeply with citizen-science projects, citizen-science activities and lessons should be framed within "multiple ways of knowing" and supplemented with "creative perspectives and approaches necessary for progress in the scientific endeavor" (115). In the complex scientific endeavor of brain science, the cognitive literary theorist Patrick Hogan grounds nebulous notions of "creative genius" in a more accessible and cogent way that this chapter argues boasts the double advantage of wedding neuroscience-oriented citizen science to the sf literature of a virtuosic writer such as Gene Wolfe. Extending the work of Howard Gardner, and complicating the Hebbian plasticity of the connectome on which *EyeWire* research depends, Hogan suggests that "to be creative, it is apparently necessary to have a hierarchy of connection strengths that does not sharply degrade remote associations, but that at least partially reflects knowledge of the relevant domain" (Hogan 122); that is, on a specifically cognitive level, mastery of a literary style or genre requires the gradual formation of domain-specific synaptic connectivity with regard to the relevant or proximal prototypes, exemplars, and schema of neuron pathways associated with that type of writing, but radical creativity and groundbreaking innovation also entail the unusual ability to make apt or recognizable cross-domain associations that are nevertheless distant and remote to the standard or conventional nodal network of connections.

Critical appreciation of Wolfe's work often comments on his febrile reworking of shopworn genre-sf tropes and icons; in the context of citizen neuroscience in the classroom, highlighting for students the adeptness and dexterity with which Wolfe synthetically fuses homages and parodies of his diverse literary influences and antecedents can dramatically illuminate the cognitive mechanisms of human creativity as a mental faculty. For instance, Wolfe scholar Peter Wright avers that "much critical opinion [on Wolfe's *Urth Cycle* of twelve novels] is divided into two mutually exclusive camps: one advocating the text as fantasy, the other perceiving it as science fiction" (56). In the effulgently creative quartet of short novels collected as his magnum opus *The Book of the New Sun* (1980–1983), attentive readers are rewarded with the realization that what initially smacks of high or epic medieval fantasy gradually reveals itself to be far-future space opera. What seems like magical wizardry turns out to be futuristic energy weapons, what immediately strikes one as castle towers actually are dilapidated spacecraft, feudal lords morph into transdimensional aliens, portals become faster-than-light time travel, soldiers are laid bare as robots, and giants exposed as genetic engineers.

In one of the most iconic descriptions in the series, in the first volume *The Shadow of the Torturer* (1980), the narrator happens upon an old picture of a "warrior of a dead world," "an armored figure standing in a desolate landscape," holding a "strange stiff banner" and wearing a gold helmet "without eye slits or ventilation" (36). This image strikes the narrator as powerfully timeless, and yet the savvy reader might have connected the clues that this picture is not of an embattled knight errant but an Apollo astronaut following a moon landing. The significance of Wolfe's unique blend of sf and fantasy here may be, as Wolfe-critic John Clute's lifelong criticism on the author consistently stresses, that Wolfe's fiction evokes a darkly ironic vision more haunted by the "underside" of sf futurism than earlier pulp, magazine, and paperback sf. In other words, as Wolfe cryptically writes in *The Sword of the Lictor* (1981), "the past cannot be found in the future where it is not" (41)—that is, under one interpretation of this line, grandly futuristic sf often erroneously recycles the threadbare archetypes and deep-seated iconography of the most escapist of historical fantasy. A cognitive literary theorist in the vein of Hogan might explain this clever synthesis of divergent genre materials as a vivid manifestation of artistic creativity precisely because it counterbalances domain-specific with domain-crossing modes of associational thinking to comment on the sf genre and our propensity for myopic and benighted future speculation more broadly.

In such a way, *The Book of the New Sun* lends itself to rich, multivalent discussions of our collective cognitive endowment by way of literary pyrotechnics that promise to be engagingly readable and intellectually fulfilling for the hailed student. The novels are first-person remembrances of the nominal protagonist Severin, a character who claims to be blessed and cursed by a preternatural memory that surpasses being merely eidetic or photographic and enters the clinical realm of hyperthymesia, a brain disorder that results from an enlarged temporal lobe and caudate nucleus. Wolfe alludes here to one of his great literary influences, Jorge Luis Borges, and his story "Funes the Memorious" (1942), which chronicles a character whose superior memory acquired from a horseback riding accident paradoxically makes him "not very good at thinking" (Borges 137). Indeed, Borges's story has turned out to be scientifically prescient; in the psychiatric literature on the effects of such excessive recall on the human mind, hyperthymesia is sometimes associated with executive-function cognitive deficits of impaired reasoning, judgment, decision-making, and abstraction. Much like the tragicomic figure of Funes, Wolfe's Severin, who begins the saga as a lowly, orphaned apprentice of the grim Torturer's Guild, has had an experience but missed its meaning, wandering incoherently in his labyrinthine memory palaces, vaporous agonized musings, and metaphysical maunderings, a dynamic testament to the palimpsestic process of the brain's connectome. Recall that the tentative, haphazard nature of the connectome encodes, retrieves, and restructures its own neural architecture, thereby continually transforming the very information it restlessly stores. In the beginning of the first book, *The Shadow of the Torturer*, Severin, mystified himself by the false perception and confused recollection of a possibly forged coin upon the acceptance of which he internally pledges loyalty to the rebel-traitor

Vodalus, frankly admits that perhaps the most "harrowing" realization of his life was that "I am in some degree insane" (27). Indeed, dedicated Wolfe readers delight in observing the endless litany of damning omissions, convenient misunderstandings, perplexing slipups, confabulating fantasias, and maddening mistakes of this self-unknowing narrator haunted with an otherwise perfect recall.

Unlike Funes, though, Severin overcomes his cognitive deficits, conflicted self-doubts, and his cruel social upbringing to assume the auspicious mantle of a Platonic philosopher-king, if only briefly; his moral and psychological arc in the series parallels a heroic Arthurian journey that ends with the deeply human, quasi-divine figure of Severin as an enlightened monarch confessing his problematic memoirs, documenting his supremely ambiguous ascent to the densely symbolic Phoenix Throne. As such a savior, the underlying recurring image of the series is Severin's role as harbinger of a rejuvenated sun, which through millennia of entropic aging and tragic hubris has terminally exhausted itself in an apocalyptic dissipation that threatens to de-aminate all life on what is now called Urth. The ending of the series promises a messianic rebirth achieved through some far-future miraculous technology of which Severin becomes a guiding vector once the newly arisen chosen one manipulates the restorative infinite energy of a collapsed star turned white hole.

Much of the immense pleasure of reading the series derives not simply from the ironic readerly detachment of condemning the arrogant or delusional narration of Severin; the enjoyment also derives from tracking what we might call the maturing bildungsroman of "emotional intelligence" in Severin, as he progresses to fulfill his epic destiny to resuscitate and enlighten a dying planet. Akin to mind-set theory, emotional intelligence is a complicated concept popularized by Daniel Goleman, but one facet of which, as John D. Mayer and Peter Salovey argue, consists of the acquired ability to reflect on, analyze, and evaluate "emotional reasoning by linking emotions to situations" (13). As we have seen earlier in this chapter, cognitive literary theorists contend that reading novels may assist with developing emotional intelligence since novel reading continually negotiates the emotional intentions, motivations, and situations of various characters, not to mention the implied author's murkily presumed intentions.

Citing Robin Dunbar's influential work that suggests the maximum cognitive load of fifth-order intentionality ("Severin thinks that Agia believes that Dorca knows that Doctor Talos understands that Baldanders thinks . . . ") becomes overtaxing in everyday life, cognitive literary theorist Blakey Vermeule argues that "Machiavellian narratives" strike readers as richly literary stimulus for such mind-reading tendencies when they feature a "mastermind" who from an Olympian position reflect "not only on the motivations of others but also on his own reflections" (87). Severin's relationships with other characters reflects the dawning enlightenment of emotional intelligence toward their recursive motivations and convoluted intentions, as the young man gradually casts off the naivety and blindness that afflicts his earlier encounters. For instance, in the first book *The Shadow of the Torturer*, when Severin originally befriends the Chatelaine Thecla while she is imprisoned as a traitress to the Autarch by the Torturer's Guild, he originally anticipates her intentions based on his torturer training, namely, that prisoners would naturally attempt to engage in conversation with their guards to bargain for their release. Thecla explicitly disowns this ulterior aim (52), even though Severin most likely doubts her sincerity. Ultimately, despite the supervisory ministrations of his guild superior Master Gurloes, Severin is indeed exiled for betraying his loyalty to the guild, sneaking in a knife with Thecla's meal so that the noblewoman can avoid a horrific torture session by committing suicide. Under one possible tabulation of the cognitive load in this situation, the nesting, regressive sequence of intentionality here is of the liminally dense fourth order that the novel soon in fact far exceeds: in their wide-ranging, extensive conversations both Severin and Thecla dance around the implicit subtext that Severin thinks that Thecla thinks that Severin thinks that she will ask him to eventually betray the guild and aid in her escape.

To focus momentarily on this one signal representative event of Thecla's suicide that haunts the whole series, the reasons Severin as incipient Machiavellian mastermind gives for aiding in this treason, as opposed to gallantly rescuing her, are more knotted and ambiguous than the relatively straightforward, though higher-order mental chess described above. Severin as narrator states his rebellion stems from an underlying admixture of emotions in that "he loved the guild and hated it too" (71) and because Thecla had patronizingly considered him a "rather a sweet boy" (80), under which jejune designation the injured masculine pride of the young man squirms. Later in the series, after, by Vodalus's urgings, Severin consumes an Alzabo feast composed of a hallucinogenic concoction of Thecla's brains and thereby psychically melds her mind with his own forever, Severin relives her suicide from her own radically divergent perspective. In *The Citadel of the Autarch* (1982), moreover, the reader finally glimpses what Severin learns in that psychic consumption: namely, that Thecla conceives of Severin as "a kindly boy who fetched books and blossoms to my cell largely because I knew him to be the last love before the doom," and she fatalistically pursues his assistance because of a "rebellion rising in me" in which "each damn herself for what she has done" (212). To recapitulate the Machiavellian compression of layered mind-reading intentionality that Wolfe encourages in the cognitive activity of his active, sustained, and critical readers: Severin aids Thecla because he both hates and loves the Guild that raised him, and he has become both infatuated with Thecla (a victim of the guild) and resentful of Thecla's patronizing estimation of him (and of his loyalty to the guild);

likewise, Thecla seeks Severin's lethal aid because she hates both the decadent state of the Commonwealth (a product of her own divided allegiance to the House Absolute and to Vodalus) and wishes to selfishly use the young Severin to sate what she recognizes to be one last erotic fling before she dies.

Wolfe balances the convoluted intricacy of such deep psychological characterization with a mesmerizing deployment of storytelling devices and techniques that promise to enthrall even the most casual or resistant reader. A study of a key White Paper sponsored by the European Commission for the purpose of advising citizen-science funders and policymakers concluded in principle that an essential stakeholder group in citizen-science projects consists of educational actors and networks who seek to motivate and empower participatory student learning through engagement with science (Richter et al. 277). It is the chief argument of this book *Citizen Science Fiction* that provoking interactional and contributory engagement with citizen science necessitates supplementing such instruction with the well-matched learning supports of sf literature and media. The dazzling cognitive overload of Wolfe's fiction provides this much-needed learning support for the otherwise abstruse matters of contemporary neuroscience.

The cognitive-literary theorist Vera Tobin, for example, connects the cognitive concept of "the curse of knowledge" to the inner workings of surprise in fiction that can be helpful in discussions of citizen neuroscience. The curse of knowledge, in cognitive-science research, refers to the immense difficulty people have discarding or overlooking information once they have acquired and internalized said data, and this paradoxical dilemma leads to numerous critical-thinking fallacies, variously categorized as hindsight bias, the spotlight effect, source-attribution errors, and the illusion of transparency. Based on recent cursed-cognition studies of stage magic and misdirection, Tobin explicates a number of literary techniques that fiction writers exploit to manufacture the effect of surprise in storytelling, including the "managed reveal," which subtly plants tantalizing hints and clues into early scenes that will be referred to and reconfigured in later scenes in such a way to "encourage us to be both forgetful and forgiving at the right times, in the right ways, to make a revelation seem especially convincing and unassailable" (103). Indeed, Wolfe scholar Joan Gordon observes that a characteristic trademark of this writer's writer is the virtuosic deftness with which he handles the carefully staged misdirection of a "plausible surprise" (46).

In *Book of the New Sun*, the number of surprising managed reveals are countless; however, one particularly gripping one in *Shadow of the Torturer* (1980) concerns Severin meeting a shopkeeper and his twin sister, Agia and Agilus. The seductive Agia originally helps Severin prepare for combat with poisonous flowers after he randomly receives a challenge from a mysterious armored officer. Cursed in this present moment by his limited knowledge—that is, on a strictly neuroscientific level—the stable cell assemblies of hard-wired synaptic chains in his connectome, as Sebastian Seung might put it, Severin speculates that the gauntlet has been thrown down because "someone—perhaps the Autarch or shadowy Father Inire—has learned the truth about Thecla's death, and now sought to destroy me without disgracing the guild" (109). It soon transpires, though, that, unbeknownst to Severin yet, in this rag-shop scene Agia has secretly and quickly costumed herself in the armor of the officer as part of a conniving plot with her brother to steal Severin's rare, expensive sword *Terminus Est*. Upon this revelation, the attentive reader is thereby enticed to reread the scene where Severin accepts the challenge and notice, for instance, that the officer's costume has a clue to whose Medusa-like face hides behind the helmet: "a golden chimera with the blank, staring face of a madwoman fluttered on its breastplate" (108). When Severin surprisingly kills a masked Agilus in the challenge, Severin cements his position as Agia's sworn enemy, and the subsequent, equally managed reveals of her surprising but inevitable recurrences throughout the series culminate with Agia finally replacing Vodalus to become the star-crossed rebel leader to Severin's newly appointed role as Autarch of the Commonwealth.

First proposed in Daniel Goleman's best-selling *Emotional Intelligence* (1995) and quickly promulgated by a bevy of business, educational, and motivational speakers, the concept of emotional fluency or competence as an empirically measurable cognitive faculty has met with strong resistance from some psychological circles. In experimental studies, recent critics of "emotional intelligence," such as Dana Joseph and Daniel Newman, counter that emotional intelligence has more lucid explanatory power when framed as a combination of general intelligence, the so-called "g factor," and its measure of aptitudes for reasoning, learning, planning, and problem-solving, as well as certain personality traits, such as low neuroticism and high conscientiousness. Nevertheless, conceived broadly and informally, discussions of emotional intelligence can be a strategically useful antidote to the rhetorical scientism of a eugenic, Social Darwinist, or biologically essentialist overemphasis on general intelligence. Beyond the management of one's own psychological well-being and self-care, Mayer and Salovey, for instance, posit that the persistent manifestation of gritty, stoic resilience—that is, the poised self-assurance associated with emotional intelligence—might very well be the key ingredient for the professional "achievement" and social-political "competency" (17) that students avowedly want from their educational experiences. In the larger context of citizen science, emotional intelligence, again conceived loosely, can be both the product of and impetus for democratic and participatory engagement with a citizen science that responds to the "underserved communities and unheard voices [which] need to be included in a people-powered science" (Nascimento et al. 233).

For many, personal emotional intelligence, not to mention the loftier ambitions of self-efficacious citizen science, is a formidably challenging goal to achieve simply in the neuroscientific terms of mental processing, neurological dynamics, and cognitive inference. Vermeule draws on the psychological experiments of Pascal Boyer to argue that the average person struggles with grasping the religious idea of a sublime, abstract divinity in part because the theistic notion entails the full-access perspectives and inconceivable intelligence of a high-Machiavellian figure who can faultlessly peer into the consciousness of all of creation with omniscient attentiveness. In a materialist neuroscientific context, Vermeule suggests, the intensely self-conscious literary writing she labels the so-called "god novel" implies the absent-present contrivances of an Olympian perspective in the fabulous zone of the fictional narrative itself, which circumscribe the limited psychological points of view of its constrained characters. A classic example of a god novel, for instance, would be James Joyce's Modernist masterpiece *Portrait of the Artist as a Young Man* (1916), in which Stephen Dedalus famously intones that the genuine artist "remains within or behind or beyond or above his handiwork, invisible, refined out of existence, indifferent, paring his fingernails" (Joyce 215). While reading the god novel in this narrow secular sense, Vermeule surmises, we "see God as personifying the collective interest, a Leviathan of obligation to whom people are willing to submit on the grounds that the costs of submitting are also shared by everyone else" (146).

An inimitable practitioner of the god novel, Gene Wolfe is fairly well-known as well for his elaborate puns on his own surname, embedding scattered references to lupine creatures in all his work. Robert Borski notes, "They're the literary equivalent of cameos, and very similar to the appearances that director Alfred Hitchcock made in his films" (141). Following the climactic battle of *The Book of the New Sun*, in *The Citadel of the Autarch*, for instance, Severin observes just such a bone-chilling cameo, when he describes the "inhuman howling" of big bad wolves feeding on the carcasses of the dead, sending him reeling into his primal childhood memories of "dire wolves in my mind, black and silent shapes, such as the onyger, pouring down into the valley" (321). Wolfe scholar Michael Andre-Driussi notes that throughout *The Book of the New Sun* Severin is often associated with the Greek god Apollo, who traditionally symbolizes art, the sun, and wolves (88). Far from merely a gimmicky gag, then, the sublime terror of this scene reverberates with a parallel scene when, later in *The Citadel of the Autarch*, Severin assumes the mantle of Autarch of the commonwealth by consuming the potent vial of a heady drug composed of the Alzabo wolf given to him by the preceding dying Autarch (Wolfe 356).

This vial endows the previously boyish and emotionally immature apprentice-torturer Severin with the regal adult maturity of his ultimate state of emotional intelligence that the book macrocosmically links to an emergent reign of peace, freedom, and social justice for the diseased, war-torn, and entropic universe at large. Upon consuming the Alzabo vial, Severin transforms into the Autarch or "self-ruler"; in this condition he regulates thousands of minds extracted from his populace and thus literally embodies the leviathan covenant of sublime obligations that constitute the collective public interest of citizens in the Commonwealth. With ironic self-reflexivity, Gene Wolfe therefore uses the sinister, rapacious wolf motif as an index of the high-Machiavellian, godlike authorial perspective that Severin has painfully attained at the precise moment when he finally embarks on a redemptive journey to ignite the new sun immediately after composing the gradually emotionally intelligent narrative we eventually read.

SHINE ON, YOU CRAZY DIAMONDS: MECHANICAL TURKS, MAPPA MUNDI, AND BRAIN JEWELRY

A significant caveat that teachers need to consider when integrating neuroscience into the writing curriculum is whether the scientific credibility invoked is overly reductive and monolithic. The critical stance of citizen science can therefore be useful in interrogating the transhistorical presumption of a basic, essential, universal, or timeless human brain existing outside cultural or political perspectives. Healthy skepticism in this regard neatly dovetails as well with a dialectical injection of sf literature and media into academic discourse, especially, as educator Daniel Ian Rubin has shown in the context of popular music, given the common cultural conjunction of Orwellian dystopia, the neurological conditioning of "thoughtcrime," and the nightmarish dark side of technoscientific discourse. In his reading curriculum, Rubin inventively pairs a close reading of Orwell's classic *1984* with exposure to accessible popular music that taps into the dystopian subgenre Orwell fostered so that his students, in their written responses, begin to register the problematic political subtext surrounding their everyday life (Rubin 77). Likewise, the landmark success of the British progressive rock band Pink Floyd and their groundbreaking album *The Dark Side of the Moon* (1973) promises to motivate students to productively question the direct applicability and easy implementation of some of the more pathologizing findings of neuroscience rampant in popular-science media.

Pink Floyd lead-guitarist David Gilmour credits the lyricist Roger Waters with creating the underlying theme of the obsessively motif-strewn concept album: "[The band] sat in a rehearsal room . . . and Roger came up with the specific idea of dealing with all the things that drive people mad" (Rose 14). From the *in utero* heartbeat of the opening track, "Speak to Me," to the last fading heartbeat requiem of the closing track, "Eclipse," the songs tackle the varied thematic terrain of being existentially adrift in time's passage to the irresistible allure of workaday capitalism. Yet the track "Brain Damage" perhaps most synopsizes the album theme of "madness" and its historical connection to the antipsychiatric movement that helped galvanize movements to deinstutionalize asylums, ban electroshock therapy and lobotomies, and influenced thinkers as varied as Michel Foucault, Thomas Szasz, and Ronald Laing, as well as psychological trends as distinct as cognitive therapy and psychopharmacology. The lyrics of this song, after all, include these harrowing lines denouncing lobotomies: "the lunatic is in my head/you raise the blade/ you make the change/ you rearrange until I'm sane."

Roy Porter suggests that "drastic splits within psychiatry as to the nature of mental illness" (4) are foundational to discourses on madness from antiquity to the present. These splits continually revolve around whether the stigma of mental illness can be recuperated to provide care, therapy, and kindness to the afflicted, or whether mental illness will continue to be used as just another politicized football to scapegoat and smite the modern equivalents of medieval heretics, the possessed, and witches. In this latter case, Sigmund Freud's earth-shattering question of *Civilization and Its Discontents* (1926) perennially arises: namely, "If civilized society is thus disordered, what right has it to pass judgment on the 'insane'?" (Porter 88) As also similarly explored in "Shine On You Crazy Diamond" from the follow-up album *Wish You Were Here* (1975), "Brain Damage" recasts the former frontman and acid-casualty Syd Barrett as the haunting specter of the band's lost innocence, a madcap holy fool chewed up by the absurd pressures of manufacturing top-charting pop concoctions on the cynical record label's dime. This autobiographical backstory, of course, mirrors in microcosm the unjust stigma of madness ritually assigned to all countercultural misfits, outcasts, and rebels lost in the Orwellian madhouse of 1970s Britain and the United States.

The last snippet of conversation Roger Waters recorded from the doorman Gerry O' Driscoll—"I've always been mad"—speaks to the audience reaction to this decades-long megaplatinum record, an audience who, as "Eclipse" dictates, collectively agree to meet up with Pink Floyd on the maligned dark side of the moon upon the record's conclusion. On this dark side of the moon, the obfuscating illusion of alternative rationalities shrouds the putative enlightenment of scientific and technological reason. Incorporating Pink Floyd's album, and its enthralling discourses on madness, into the classroom, however, does not imply that the pedagogical goal of neuroscientific literacy has to be utterly dismissed or repudiated. The neuroscience of music also holds dynamic potential for the nuanced incorporation of Pink Floyd's sf dystopia into the science-oriented writing classroom.

In *Sweet Anticipation,* David Huron, for instance, theorizes that music derives its emotional power from disrupting cultural expectations and the subsequent neurological combination of fast reaction response and slow appraisal responses triggered by the limbic system and the flight-or-fight mechanism of the brain region called the amygdala (Huron 16). Pink Floyd's *The Dark Side of the Moon* offers attentive student listeners a gamut of neurological surprises ripe for close analysis and evaluation in written compositions, from the wailing hurricane of Gospel singer Clare Torry's nonlexical vocals to Richard Wright's jazzy piano chords, from the innovative experiments of the VCS3 synthesizer to the irregular time signatures, atonal found noises, instrumental interludes, and endless improvisational grooves. Evoked as well by Storm Thorgerson's iconic album cover art of an abstract Newtonian diamond refracting a concrete rainbow spectrum as much as the renowned concert magic of laser light shows and assorted bombastic pageantry, the psychedelic vibes associated with Pink Floyd's space-age musicality cements its emotional impact only when juxtaposed with the recognizable conventionality of David Gilmour's melodic blues riffs, the delicate mix-tracks of sweet harmonies, the dissonant fuzz of the rock guitars, and the professional pop song structure.

Since the 1970s heyday of Pink Floyd's popularity, the suspicion of psychological science has mushroomed beyond critiques of the mental-health profession and pharmaceutical industry. In recent years, psychology has been embroiled in what is often called "the replication crisis." In 2005, John Ioannidis published an influential article, "Why Most Published Research Findings Are False," that claimed 40 percent of published studies in the medical fields are unverifiable, if not exactly empirically wrong; likewise, in a study of cancer research, Amgen failed to reproduce the results of forty-seven out of fifty-three published papers. After a 2011 report was published in *Journal of Personality and Social Psychology* that claimed to find evidence for paranormal premonitions, this replication crisis has frequently been associated with psychological science in particular. In a survey of the problem, Scott Lilienfeld and Irwin Waldman posit that the replication crisis can be explained by myriad factors, including the following prevailing yet dubious scientific practices: the decline effect in which scientific claims become less defended over time, which can be partly accounted for by the lack of splashy publishing incentives for confirmatory studies; the reverse inference or potential logical fallacy of reasoning backward from brain activation to cognitive function in brain-imaging studies; the cherry-picking confirmation bias of selectively seeking specific evidence and discarding inconvenient results; the lack of sufficient controls or other experimental ballast; the allegiance effect of being biased by expertise; the problem of Hypothesizing after Results are Known (HARK-ing), and overhyping or the "p-hacking" of statistically insignificant results. Rather than encouraging a wholesale debunking of psychological knowledge that may induce apathy or resistance toward psychology among uninitiated students, though, reviewing such strenuous objections can more productively serve to highlight the uncertainty and doubt-laden limits of even sound scientific rhetoric and research in a fashion that challenges the authoritarian credibility the public too easily confers to the official reception of popular science.

Likewise, while the replication crisis rages, the crowd-sourcing platform of Amazon's Mechanical Turk (MTurk) service has injected renewed energy into psychological experimentation in a way that lends itself to a thoughtful, analytic, and reflective writing assignment. MTurk is a remote contracting business that employs hundreds of thousands of workers competing for a few cents on microtasks called Human Intelligence Tasks (HITs), such as completing surveys or responding to questionnaires about visual or verbal content. While its labor-compensation practices and its exploitative endorsement of the gig economy have been justly and widely criticized, MTurk has also been instrumental in the scientific production of thousands of studies published in prestigious journals every year, with psychologists especially claiming that the service may "reduce the biases found in traditional samples" (Buhrmester, Kwang, and Gosling 4). In contrast, other researchers believe MTurk exacerbates the replication crisis, finding "compelling evidence of a decrease in MTurk data quality" (Chmielewski and Kucker) and raising worries over the rise of "bot farms" or online marketplaces that sell software that can cheat the system with automated programs that imitate workers. Akin to Edgar Allen Poe's namesake Mechanical Turk, Amazon's MTurk may fast become more of an ingeniously lucrative hoax and uncanny publicity stunt than the long-anticipated harbinger of a rational techno-fix to the replication crisis.

Whether or not citizen-science-adjacent projects such as Amazon MTurk will tend to become a symptom or cure of the replication crisis, probing the problem in psychology additionally interrogates some of the core presumptions of neuroscientific research, such as the high-flown ambition to map the neural net of the human connectome. Examples abound in sf media and literature that depict neuro-augmented posthumans as marvelous objects of sublime beauty and terror: analyzing these connectome fictions in the writing curriculum can serve to underscore the promises and perils of psychology as an epistemic discipline for which the average citizen invests as a stakeholder or empowered critic. Comparing the medieval allegorical maps of the world to the contemporary neuroscientific quest to chart neural nets, Justina Robson's *Mappa Mundi* (2006), for instance, tracks the aftermath of the invention of a brain simulation technology, called NervePath in the novel, which then loops back to rewire and upgrade the chemical-organic architecture of the brain itself. Given speculative leaps in nanobot technology, NervePath promises to finally achieve the dream of tracing a neural net, dethroning the metaphysical seat of the soul into an entity entirely "measurable, definable, mappable" (115) by the scientist-hero Natalie Armstrong. In her imaginary extrapolation on current neuroscience, Robson adapts Richard

Dawkins's influential notion from *The Selfish Gene* (1976) of sociobiological replicating memes that vary, evolve, and persist by mechanisms of cultural selection analogous to Darwinian natural selection. The mappable consciousness of an individual's neural net forms a "memeplex" through which one "unconsciously filtered all knowledge of the world."

According to Natalie Armstrong's vision, the NervePath mind-mapping can potentially be harnessed and reprogrammed to eradicate hierarchical, discriminatory, and exclusionary ideological memeplexes such as "racism, a conditioned social reaction" (382). However, with the cloak-and-dagger involvement of a shadowy global cabal of espionage and conspiracy, the discovery soon becomes weaponized and threatens to instrumentalize brains of dissidents into disposable machines, converting agency into "slavery at the hands of cruel masters who enjoy the depravity that their technology is able to bring" (521). Against the connectionist assumption that the mind can be mapped with excitatory or inhibitory neural threads standing in for weighted values in a computer algorithm, Robson echoes a highly speculative quantum theory of consciousness that has affinity with Roger Penrose's *The Emperor's New Mind* (1989) that hypothesizes the indeterminacy of quantum mechanics renders the inner workings of the mind incalculable, noncomputable, unmappable. This wildly speculative gesture reinforces the flaws, uncertainties, and limits that underpin neuroscience in a likeminded skeptical vein as that of contemporary critiques of the replication crisis.

As a character infected by NervePath explains, "Matter . . . is energy plus information," and therefore the mind enhanced by new viral memeplexes fantastically manipulates physical interactions at the quantum level, securing a transformative posthuman space for individual autonomy, unpredictability, and randomness to reign unimpeded. Another character, Mikhail Guskov, a globe-trotting shapeshifting superspy turned radical psychologist, suggests that "emotion was the master switch" (447) for mind-mapping such that if you alter "the emotional portrait of a meme and you alter identity without the need for fancy fiddling with the hugely tangled and difficult definitions of neural pattern and synaptic timing" (448). Yet, as Natalie realizes, once she too becomes infected by NervePath and phase shifts into a disembodied quantum posthuman, this emotion master switch acts as a gateway to transcendent communion with understanding the mind in all its irreducible, fractal complexity. Before she willingly sublimates her own physical substance into an ethereal form, in the spirit of open-source dissemination of scientific knowledge, Natalie decides to release the world-changing emotional experience of this mind-mapping technological novum to the public.

Even though at first blush the specialist domain of cognitive science may seem only to be especially relevant to training future expert neuroscientists and medical professionals, in recent years, as citizen science has risen in prominence, so too has cognitive science become more explicitly geared toward studying emotional expression and therefore widely adopted in education as well, substantiating theories for how the brain learns through emotional response, articulation, and regulation. As Kathleen McCormick argues, contemporary educational discourses, by updating traditions that date at least back to Rousseau and Dewey, have seized on the master switch of the emotional background experience of the student to push for "expressivist" schools of reading and writing pedagogy that has "revitalized the teaching of literature" (40), especially in its active resistance to the deficit and banking model—in which a student is a passive, empty vessel waiting to be filled by direct instruction—prevalent in some interpretations of "cognitivist" theories of learning. McCormick argues, however, that such expressivist schools are ultimately as incomplete as cognitivist theories since they also dubiously promote a nebulous, romanticized vision of individualistic subjectivity in the process of challenging the objective science of the rational-logical brain. McCormick's binary between "cognitivist" and "expressivist" pedagogy, though, does not seem as relevant today as in previous decades, as these two schools of thought seem to be increasingly hybridized and intertwined in recent neuroscience. Nevertheless, McCormick's underlying critique retains its forcefulness against this hybrid version of cognitivist-expressivist pedagogy when such theories may be understood to naively posit that students live in a social-cultural vacuum and engage only in rational or emotional cognitive discourse. Enshrined with the aura of cognitive science, cavalier positions that suggest this dehistoricized conception of the inner workings of the brain oversimplify the complex richness of a student's entanglement and positioning in contemporary discourses of science and technology. For instance, popular mythologies, misconceptions, exaggerations, and even straightforward junk science have become widely disseminated culturally as self-legitimating and self-evident staples for how the brain ostensibly works.

The independent videogame *Gray Matter* (2010), designed by Jane Jensen, and released by her Pinkerton Road Studios, immerses players in the process of sifting through pseudoscience, credible science, and fringe science within the role-playing adventure framework of a gothic detective narrative and gameplay. In artfully rendered animation, the players control Sam Everett, a plucky, problem-solving young female magician masquerading as the assistant to the Oxford neuroscientist David Styles, a masked and brooding man in the vein of *The Phantom of the Opera,* as she investigates the seemingly paranormal events that swirl around the death of his wife and the conducting of his bizarre brain-imaging experiments. In a dynamic merging of so-called "female gothic" and "male gothic" traditions, the mad experiments are revealed to be nonoccult, if highly unorthodox, and at the same time a distraught student is revealed to have murdered David's wife and perpetuated a hoax of ghostly visitations through her family curse of burgeoning telekinetic powers. Meticulously reconstructing this surprising reveal through newspaper microfiche, magazine clippings, interviews, letters, lab notes, riddles, magic tricks, mazes, and other enjoyable and cleverly sleuthed physical clues and retrospectively imagined events, players must take on the persona of the theatrical performer who creatively deduces for herself the stage magic occurring beneath the distracting gothic paraphernalia of spectral hauntings. In the process, players endeavor to distinguish between more or less credible and ongoing areas of neuroscientific research on fugue states, comas, hallucinations, altered states of consciousness, and the electro-chemical basis of neuronal firing from supernatural and fantastic phenomena of ghostly visitations, demonic possessions, pyrokinesis, and communing seances with the recently deceased. Upon finishing this intricate game, students may also acquire a newfound respect not only for the genre trappings of such sf-induced horror fiction but for the idea that New Age superstition may function psychologically as wishful projections of the grieving process.

Educators informed by these recent advances in cognitive science look to such research to validate their views of reading and writing as constructing unique and fluid bases of neurobiological development. As a result of this widespread recourse to cognitive and neuroscientific sciences in pedagogical theory, what once was basically conceived as the eternal, unchanging Platonic essence of student knowledge has transformed into a plastic palimpsest of mental processes that continually collate emotions, memories, ideas, and experiences. This malleable and dynamic vision of learning emphasizes how long-term storage and processing of knowledge depends on a dynamic convergence zone between the new pieces of knowledge to be acquired, self-concepts based on past experiences, and an evolving set of cognitive belief systems informed by cultural and social contexts. Thus, as David A. Sousa reminds us, the brain, in the last analysis, bears no resemblance to the banking or deficit model of the public understanding of science, which entails static, fixed notions of one-dimensional information processing. By contrast, the reception of scientific knowledge by students should be viewed as an "open, parallel-processing system . . . [that] analyzes, integrates, and synthesizes information and abstracts generalities" (39) from a changeable and dispersed variety of circuits, inputs, and feedback. With this fluid reconfiguration of the mind now widely accepted, grappling with the implications of cognitive science for citizenship and education has therefore become an increasingly important task for writing pedagogy.

Martha Nussbaum traces the citizenship rhetoric of modern education back to ancient Stoic philosophy, arguing that Stoic-influenced Socrates, Seneca, and Marcus Aurelius in particular set the Western template for viewing institutionalized learning as a probing, argument-driven self-examination unclouded by the fact that "public life is frequently rendered irrational by the power of sentiments such as anger, fear, and envy" (29). In other words, emotional intelligence does not repress, diminish, or discard the inherent value and validity of a student's rich, unique emotional mindscape as an essential cognitive tool that enhances the learning process. However, the goal of developing keen emotional intelligence as a signal aim of higher education does endeavor to regulate, manage, or control the messy, complicated emotions that lack sufficient reason-driven motivation or justification to be continually encouraged in a cohesive civil society. While evaluating the future trajectories of present-day citizen-science coups such as *EyeWire* or the Mechanical Turk, a student may critically reflect on citizen cognitive science as a mode of facilitating the emotional regulation of received public opinion as much as it can challenge the expertise-laden biases of the replication crisis in the psychological sciences. Likewise, to this end, the remarkably hard-sf writer Greg Egan has written a number what we might for convenience label "connectome fictions" that speculate on the future of mind uploads and advocate for the paradigm-shifting power of neuro-literacy that equally scrutinize the desirability of brain-mapping, let alone brain-augmenting projects.

In the masterly short-story "Closer," for instance, Egan charts out the rippling consequences of a mind-upload device called a "jewel" that in a future society is implanted in children at birth and then consensually "switched" with one's organic brain when an individual turns eighteen. The masculine narrator, Michael, is especially enamored of the transition of his mind to this naturalized technology that runs on a faster, more durable, and more flexible computational substrate: "My brain was a kind of bootstrap device, nothing more, and to mourn its loss would have been as absurd as mourning my emergence from some primitive stage of embryological neural development" (Egan 310). Yet perhaps by dint of the gadget fetishism of the jewel device, the narrator is also obsessed with the metaphysical problem of solipsism, and when he falls in love with an exciting woman, Sian, they both decide to participate in a number of radical experiments to feel what it is like to be psychologically closer to each other—they swap bodies, genders, implant their jewels into clones, and even temporarily merge minds into a single entity. This last experiment destroys the romantic glue that attracts the two to each other in the first place, effectively breaking the couple up since it eliminates the element of "the alien, the unknowable, the mysterious, the opaque" (312) and the existential basis of their emotional connection. In this extraordinarily teachable story, Egan constructs a touching parable about the misguided, counterproductive conflation of high-handed neuroscientific experimentation with a more nuanced understanding of lived emotional intelligence, underscoring the unbridgeable gap between our limited scientific knowledge of the brain's mappable connectome and the irreplaceable, direct human experience of the mind's qualia whose very ineffable alterity forges vital, mysterious bonds between divergent selfhoods.

The popular trope of the qualia-leached, anhedonic mind upload typically lends itself to questions and debates surrounding the neuroscientific nature of disembodied emotions since, as Kathleen Ann Goonan frames this "conceptual pudding," "complexly linked biological systems generate emotions"; hence we can only presume that "emotional development occurs in tandem with our physical growth" (197). To be sure, Egan can be eminently more sanguine about the prospect of boosting civic duty through augmenting neural hardware. At the end of the novel *Permutation City* (1994), for instance, two mind uploads, called "Copies," Paul Durham and Maria Deluca, are debating in a virtual realm known as "the Autoverse" about whether to construct another artificial world-within-a-world after the previous one abruptly implodes. Beset by existential ennui, Paul wants to suicidally quit his artificial program forever; however, at the behest of Maria, Paul simply adjusts his "exoself," or his inner program's parameters, to physically want to persist with the next daunting stage of rational scientific pursuit. New to the Autoverse herself, Maria marvels at the majesty of this literalized emotional regulation and the extreme challenges it poses to the conventional notion that individuals are merely passive flotsam and jetsam at the mercy of their deeper emotional undercurrents and drives: "Had he granted himself transhuman resilience, and healed himself of his terminal despair . . . or had he died in silence, beyond her sight, and given birth to a companion for her, a software child who'd merely inherited its father's memories?" (309). Whether viewed as fantastic wish-fulfillment or plausible extrapolation on recent neuroscientific breakthroughs, Egan's mind uploads unsettlingly update the stoic discipline of emotional intelligence for the transhuman future.

In *Zendegi* (2010), Egan, though, qualifies his otherwise ebullient endorsement of neuroscience by way of directly commenting on the limits of Human Connectome Project (HCP), which, as mentioned at the start of this chapter, has a significant citizen-science component and has inspired the gamified research of *EyeWire*. In this novel, the Iranian scientist, Nasim Golestani works for the HCP and maps the avian brain of zebra finches to produce original computer simulations of their birdsong. However, the HCP does not ultimately pursue Nasim's research, freeing her up in a future timeline to create virtual uploads of human minds on her own. Written decades after *Permutation City*, *Zendegi* is much more circumspect about what can reasonably be expected from the technoscientific prospect of mind uploads. When Nasim is hired by the terminally ill Martin Seymour to create a mind upload—or "proxy" in this novel—of his personality to provide guidance and comfort to his son, Javeed, upon Martin's imminent passing, Nasim valiantly fails at such a heroic scientific effort. At the end of the novel, Nasim indulges in a bittersweet meditation on the quixotic nature of her bold attempt: "Whether or not [mind uploads] proved to be possible, it was a noble aspiration. But to squeeze some abridged, mutilated person through the first available aperture was not" (Egan 329). Egan's transcendent transhuman vision of digital immortality has been postponed, if not permanently curtailed, yet the functional mapping of the brain, and the full flowering of its ilk of assorted citizen-science projects, are loosely affiliated with a revolutionary overthrow of Iranian theocracy and decidedly not in league with the so-called "cis-humanist" tactics of a radical terrorist fringe that attempts to disrupt the technological development of mind uploads.

A pedgagogical focus on emotions in cognitive science, and its provocative intersections with the AI research discussed in chapter 3, may yield mixed and fruitful resources for writing-oriented discussions and assignments, as can be readily seen, for instance, by the incorporation of *Distress* (1995) by Greg Egan into a citizen-science pedagogy. The novel concerns an investigative video journalist who visits the anarchic-socialist utopia called Stateless founded by a politically engaged population of renegade scientists. At one point, the protagonist has been infected with cholera as collateral damage of an assassination attempt on a Nobel Prize physicist who is planning on immigrating to Stateless. Being treated for the infection and watching the visualization of the cognitively rewiring miracle drug that will cure him at work on a digital tablet, the protagonist lashes out against his aggressors in favor of a fervently rational-empirical vision of the material universe: "I have been tempted by the Ignorance Cults—and maybe I half understood what drove them, now—but what did they have to offer in the end? Alienation from reality. The universe as an unspeakable horror endlessly denied, shrouded in saccharine artificial mysteries, every truth subjugated to doublethink and fairy tales" (Egan 262). Walking students through a discussion of the complex twists and wrinkles of this unfolding argument—in all its fullness that encompasses both the misleading and naive ideologies of authoritative scientism as well as the the posttruth disinformation of antiscience actors in *Distress*—can be a quite provocative way to unpack the extent to which the prestigious authority of expert cognitive scientists, amplified by the power of citizen-science projects, both challenge and reinforce the scientific illiteracy promoted by the Ignorance Cults.

After all, Egan evenhandedly probes the nuanced motivations of the scientifically illiterate Mystical Renaissance group; likewise, working cognitive scientists both in education and elsewhere in culture regularly advocate the startling relevance of the latest cognitive-science findings without enough scrupulous attention to the social and political uncertainties inherent in such enterprises. In *Distress*, Egan takes the thought experiment one step further and suggests brain-science research is deeply entwined with other cutting-edge bioengineering investments, and even the recondite theoretical-physics search for the Theory of Everything. Students who read this novel critically might investigate the dangers and values of cognitive science as a discipline caught in the replication crisis, and how scientific pursuits at times entail problematic risks for society at large that may or may not be worth the theoretical gamble. As fodder for such vigorous debate, Greg Egan's fiction toes the rhetorical fine line between debunked pseudoscience, transformative citizen science, and traditional elite science that Stephen Jay Gould, in *The Mismeasure of Man*—while challenging the scientific racism of Intelligence Quotients that gave rise to the much more reputable cognitive theory of general intelligence (the g factor)—famously refers to as "biological determinism." This distinction between biological determinism and more socially aware varieties of contemporary cognitive science constitutes a complex border zone between citizenship and science that can be a fruitful and rewarding arena for student inquiry.

In a way consonant with but more immediately accessible than Egan's learned fiction, the musical artist Grimes could also furnish a useful cultural touchstone for students seeking to register the impact of cognitive science on the popular imagination as well as their own emotional experience. In the hit song "We Appreciate Power" (2018), from the Japanese release of the album *Miss Anthropocene*, for instance, Grimes pairs heavy distorted guitar loops and reverbed percussion with sweeping synthesized scores overlaid with ethereal vocals that can be read as misanthropic, transhuman propaganda in favor of an imminent AI singularity. The high-pitched lyrics drone: "And if you long to never die/Baby, plug in, upload your mind/Come on, you're not even alive/If you're not backed up, backed up, backed up on a drive." Implicit in these lyrics is a half-facetious, half-serious allusion to Roko's Basilisk, or the internet-generated paradox that a malevolent artificial intelligence could in the future torture any dissidents who failed to usher in its emergence, and therefore those who are aware of this imminent apocalyptic possibility are inherently cursed by this foreknowledge to be treated as a said tortured dissenter. Grimes extends and complicates this entertaining paradox in light of the heated sf-nal speculation that often swirls around the digital immortality of backed-up mind uploads, punning that one is not alive, as in one is stimulated, unless cybernetically enhanced, and therefore, more menacingly, perhaps no longer alive biologically at all. Likewise, the subsequent lyrics, "biology is superficial/intelligence is artificial" upend the biological determinism that the varying calls to map the connectome, create neural nets, and make pedagogy more neuroscience-informed implicitly rely on. The head-banging pop hooks of Grimes's equally nightmarish and optimistic song derives from its speculating that a future computer simulation may not only supplant but supersede the limited, superficial cognitive potential of the human brain. Yet the abject blind faith in inevitable technological acceleration needed to accept the extreme premises of Roko's Basilisk, or the marketable need for such a catchy pop song to function as future agit-prop, serves as a potent reminder of the basic emotional frailty and vulnerability of the human mind that far surpasses the supremely logical machinations of purely computational modes of cognition.

FIVE

Contemplating the Science-Fictional Cosmos

Citizen Science, Astronomy, and Mindfulness Rhetoric and Pedagogy

CITIZEN-SCIENCE PROJECTS AND SERVICES

Galaxy Zoo
Planet Hunters
The African Cultural Astronomy Project
SETI@home
StarDust@home

SF LITERATURE AND READING

The Hainish Cycle, by Ursula K. Le Guin
The Binti Series, by Nnedi Okorafor
Ball Lightning, by Cixin Liu
The Orthogonal Trilogy, by Greg Egan

SF MEDIA AND CULTURE

Black Panther: The Intergalactic Empire of Wakanda, by Ta-Nehisi Coates
The Mothership Connection, by Parliament-Funkadelic
The Expanse, by James S. A. Corey and Naren Shankar
The Last Jedi, by Rian Johnson
Mass Effect, by BioWare

In 2007, astrophysicists Chris Lintott and Kevin Schawinski founded Galaxy Zoo, a wildly successful citizen-science project that initially used open-sourced data collected from a large wide-angled photometric telescope called the Sloan Digital Sky Survey (SDSS) to observe, classify, and analyze a proliferating variety of observed astronomical phenomena. In only the first phase of the project—which has since diversified into over fifteen phases in as many years—150,000 volunteers completed over fifty million classifications. In *The Crowd and the Cosmos* (2019), Lintott describes Galaxy Zoo as meeting the sorely needed demand of the current field of observational astronomy for which surging technological and scientific advances have led to an overwhelming data deluge, an embarrassment of informational riches that has become a disciplinary dilemma of identity-crisis proportions. Lintott contends that "the rich resource [of the SDSS] opens up the prospect of new insights into the processes of stellar evolution, but they also make the challenge of data-driven science apparent" (31).

The inconceivably immense vastness of cosmological data available to scientific observation has always been an unsettling anxiety of astronomy from William Herschel's eighteenth-century study of globular clusters to Edwin Hubble's epochal work at the Mount Wilson Observatory, in 1924, that provided conclusive evidence for the enormous scale of the expanding universe. Indeed, deputizing legions of interested amateurs in closely and systematically observing the staggering infinity of starfields visible from the Earth's surface, or more recently via satellite telescopes, has continually been a chief feature of astronomy, leading Martin Harwit, in *Cosmic Discovery* (1984), to identify one of the major engines of discovery in the field as a whole, alongside technological innovations, military applications, and historical accident, to be the original contributions of adept amateurs on the marginal "outskirts" (20). Yet to a significant degree the myriad affiliated citizen-science projects of Galaxy Zoo—collectively known as the Zooniverse—have shaken up the traditional field of astronomy, galvanizing recent energetic debates and research, such as those generated over the so-called "Voorwerps," discovered by the Dutch amateur-astronomer and Zooite schoolteacher Hanny van Arkel, or the classification of the specific subtype of starburst galaxies named Green Peas. The Voorwerp, a blue blob in the sky, has especially catalyzed professional academic discussion, as scientists weigh the competing probabilities over whether the observed astronomical phenomenon can be understood as a dead quasar residually shining in a dwarf galaxy or a slender cone of material jetting from a supermassive black hole.

Citizen science that harnesses the datasets, excitement, and insights of amateur astronomers helps bridge the otherwise widening fissure between science and the public. Naturally, scientific experts frame, filter, and buffer these civic contributions within the constrained institutional context of established scientific procedures, equipment, and systems of training and quality control, not to mention the conventional academic protocols of professional publication, grants, conferences, and jobs. Nevertheless, when a writing instructor carefully incorporates Zooniverse research into the composition curriculum, citizen science may serve both as primary and supplemental reading material that ideally percolates active classroom discussion and interrogates the significance of science as a societal and cultural force. Productive problems regarding existing scientific uncertainties, persistent questions, and ongoing arguments can then become fruitful focal points for the contemplative student fascinated by astronomical issues but who may otherwise feel alienated by some of the abstract, inaccessible, and remote knowledge inculcated in their science courses.

In writing pedagogy, in part to combat media-saturated attention fatigue and posttruth miseducation campaigns, there has been a restored interest in teaching mindfulness as a sophisticated rhetorical practice. Renea Frey couples the embodied physicality of such a stance of deliberate meditativeness and calm listening with a nuanced resistance to the dominant abstract logic of technoscientific rationalism and aggressive individualism. Frey argues, "by linking deep listening and empathy to practices that encourage conscious, embodied awareness, we can support students in inventing more ethical, effective arguments that address the exigencies we face in our interconnected, but precariously endangered, world" (92). Likewise, in "A Brief History of the Current Re-emergence of Contemplative Education," Patricia Fay Morgan canvasses what she terms the contemporary "third wave" (208) of mindfulness pedagogy as an intricate historical confluence of cultural, psychological, and neuroscientific discourses and rhetoric, ranging from the widespread global resurgence of popular fascination with Buddhism and Yoga, to the physiological recommendations of sports medicine, to psychological studies that suggest meditation combats depression and anxiety and aids attention control and emotional regulation, to brain scans that reveal meditation practices restructure the connectome, activating "subcortical grey and white matter, brain stem and cerebellum, [and] suggesting that the effects of meditation might involve large-scale brain networks" (Tang et al.).

As suggested by these neuroscientific findings, perhaps the demarcating line between abstract scientific and empirical rigor and the orderly, logical discussion of essayistic writing as separate and distinct from mindful embodiment and emotional self-awareness should not be so finely drawn. Moreover, pedagogical calls for teacherly investment in contemplative rhetoric and meditation often overlap with recent civility debates in composition and rhetoric studies more broadly. In a reflective inquiry into what he calls the "neoclassical approach" evident in introductory undergraduate rhetoric-and-composition textbooks, Craig Rood characterizes the civility debate as often pivoting between pro-civility proponents of invitational arguments and anticivility advocates for confrontational positions. On a higher-order level of generalization, Rood's own championing of an embodied civic discourse, or "situated civility," therefore epitomizes the mindfulness approach to civility and democratic rhetoric, balancing the competing claims of the opposed camps. On the one hand, Rood supports the adherents of invitational pro-civility argumentation and their championing of "non-assertive rhetorical practices like silence, delaying, reflection, and sincere questioning" (341). On the other hand, Rood defends the anticivility apologists of "radical speech or writing where students are instructed on forming coalitions, rallying followers, and pressuring the opposition" (342). Indeed, this holistic and balanced approach of situated civility deeply resonates with the thesis of this chapter that offers suggestions and strategies for incorporating the citizen-science rhetoric of the Zooniverse, and other astronomical citizen-science projects, into the writing classroom. Instructors can greatly benefit from marshaling citizen-science projects in conjunction with thrillingly fascinating works of sf literature and media to urge students, as empowered instruments of direct democratic action, to participate readily in the coproduction of the scientific knowledge that constitutes the cosmological imagination, while simultaneously inviting contemplative writing on astronomical discourse that is more "exploratory, hesitant, nuanced, open-ended, uncertain, or questioning" (339).

A civic-minded struggling with the uncertainties and limits of existing scientific knowledge opened by the most recent astronomical data deluge has formed the aligned cornerstones of both citizen astronomy and the perennially popular science fiction subgenre loosely known as space opera. One of the major initial astronomical discoveries ensured by the citizen contributors to Galaxy Zoo, for instance, entails the overturning of the conventional wisdom that young blue galaxies are necessarily spiral shaped and old red galaxies need to be elliptically shaped given the mass civic observations of red spirals and blue ellipticals. Galaxy Zoo was also instrumental in showing that red spirals in particular have a relatively high percentage of barred-disk formation patterns, suggesting that these galaxies mature faster than their nonbarred, lower-mass counterparts. These important empirical scientific contributions that have broad ramifications for the understanding of our own galactic futures helped to awaken the astronomical community at large to the power of citizen science—at its peak reliably classifying seventy thousand galaxies per hour—such that Michael Nielsen can today noncontroversially argue that "Galaxy Zoo is becoming a general-purpose platform connecting professional astronomers to interested members of the general public, so they can do science together" (142). The high volume of information and corresponding spike in cognitive resources increasingly available in observational astronomy challenges assumptions and commonplaces that underpin the established pieties of the field, such as the ones mentioned above connected to modeling principles for incipient galaxy formation.

A scrupulous composition instructor may reasonably raise concern over why a student would be even mildly curious about such seemingly abstruse debates. However, as educational studies of Galaxy Zoo indicate, participant curiosity, engagement, and wonder may be initially motivated by the imagination-stretching "vastness" (Raddick et al.) of the astronomical picture of the universe as well as the basic beauty and terror of cosmic scale. Such initial engagement may then develop into a more complicated understanding of deeper astronomical erudition and even eventually contribute to scientific research. Likewise, to assist in fertilizing motivation to write passionately, an instructor would be wise to consult sf literature and media as ready-made tools for appreciating the astronomical sublime. In particular, the sheer ineffable scope and intricacy of the post-Hubble conception of the universe has been a consistent, enduring feature of space opera, a sf subgenre solely dedicated to popularizing interest in cosmic vastness. In *Foundation's Edge* (1982), for instance, Isaac Asimov captures the magnificent grandeur of cosmic scale exhibited even by our own singular galaxy alone when the radical Trantorian councilman Golan Trevize explains to the cloistered scholar Janov Pelorat by way of an immersive holographic simulation of the multi-armed spiral Milky Way galaxy, which in Asimov's future history has been populated by seven-thousand inhabited star systems slowly colonized over a span of twenty thousand years: "It spread out like a gigantic, glowing whirlpool, with curves of darkness, and knots of brightness, and a central all-but-featureless blaze" (79). At this point in Asimov's sprawling saga, Trevise has only just begun his desperate mission to rescue human civilization from an abiding rut of barbarity by rediscovering the long-lost planet Earth, the delicate, rare origin of all sentient life and a near-mythical needle in the galactic haystack. The cognitive estrangement inherent in this extravagant cosmic vista is visually evoked as well by the memorable Del Rey paperback cover of *Foundation's Edge* illustrated by Michael Whelan of a prototypical sf Rückenfigur, that is, a silhouetted individual seen from behind, atop a ruin of classical pillars, dwarfed by and gazing fixedly at what appears to be a grand-design spiral galaxy.

Even in its most outrageously escapist-fantastic varieties, space opera regularly elicits mindful contemplation of the immeasurable immensities of the cosmos. In space-operatic popular culture, scientific visions of the cosmos ranging from *Star Trek* to Carl Sagan, and from *Nova* documentaries to the Marvel Cinematic Universe, frequently depict the revelatory adventure of space exploration as a contemplative outgrowth of ancient human yearnings as much as humanity-saving riposte to the pork, warmongering, and elitism of entrenched science; as Karen Schroeder Sorenson limns the rhetorical logic behind such astronomical popularization, this cosmic adventure rhetoric entails a mission of the unconquerable human spirit, the spiritual dimension of Carl Sagan's oft-quoted biochemical "star stuff," cultivating a messianic oneness with a self-aware universe: "If humans are consubstantial with the universe, then they have the power to control or influence creation" (Sorenson 116). Indeed, one of the enduring appeals of the *Star Wars* franchise of media spectacle, to its vast and vibrant fandom, consists in the visually stunning evocations of mindful atonement with the Force, a vital cosmic magic of grand destiny courted by both Jedi and Sith apprentices, knights, and masters alike, which is also often linked to disciplined meditation practices.

In the film *The Last Jedi* (2017), for instance, the protagonist of the sequel trilogy, Rey (Daisy Ridley), travels to the remote planet Ahch-To to recruit the hermit Luke Skywalker (Mark Hamill) in the rag-tag, underdog struggle against the re-emergent fascist empire of the First Order and its iron rule overseen by Supreme Leader Snoke (Andy Serkis) and Luke's former apprentice, Kylo Ren (Adam Driver), who has mysteriously turned to the dark side of the Force. A reclusive self-exile who has abandoned his ambition to create a Jedi Academy, the aged Luke reluctantly teaches Rey to practice cross-legged seated meditation and become sensitive to reconciling the harmonic balance between the light and dark sides of the Force in her journey to become a jedi knight. In pleading tones, Luke urges Rey to psychologically absorb the tranquil stillness that rests at the holy center of all the exuberant, fluctuating energy of the living universe. This Lotus-position seated meditation recurs again later in the film when Luke resumes meditation to attempt to lure Kylo Ren back to the light side, correcting the wayward apprentice's own spiritual and psychological misalignment, while also providing a distraction for Rey and the Resistance to escape an invasion. In a Force manifestation of his own remotely projected image, Luke, meditating from across the galaxy, stands off against Kylo before a fleet of gigantic Walker war machines, as they imperially lay siege with shock and awe to a Resistance base.

Both of these training sessions immediately fail as Rey and Kylo are too seduced by the raw, chaotic power of the dark side to achieve the serene, nonattached mindfulness that Jedi Master Luke seeks his apprentices to embrace. Yet Luke vanishes into a Force ghost at the end of the film, exhausted off the mortal coil by his virtual encounter with Kylo Ren, while seated before a cloudy binary sunset on Ahch-To, over the soaring crescendo of John Williams's "Luke's Theme" in the score. This transcendence implies that despite his manifold failures and shortcomings Luke has attained a peaceful, confident wisdom and enlightened aplomb by attending to the multifaceted yearnings and impulses of the cosmos. Luke's death scene visually echoes, of course, the iconic cinematography from the original *Star Wars: A New Hope* (1977) that frames the post-adolescent Luke, in Rückenfigur, an archetypal fairy-tale symbol of tyrannized children, scanning the horizon of a binary sunset on his home planet of Tatooine, an unlikely peasant farmer on the cusp of a legendary heroic journey. And indeed the enigmatic coda of *The Last Jedi* once again reimagines this scene at the end of the film, suggesting a renewed cosmic wonder, mystery, and re-enchantment in the face of the stultifying tedium of a strictly secular-material scientific worldview, as a drudgery-subjected orphan boy, wearing a Resistance ring, once again in Rückenfigur, uses the Force to telekinetically grab a broom and stare dramatically upward at a ship jumping to hyperspace, suggesting that a nostalgic invisible republic of dissidents on the uncivilized outskirts will forever forcibly challenge vast draconian empires bound to collapse with the ravages of cosmic time.

In addition to the pedagogical expediency of tapping into such pop-culture mythos and lore, integrating relatively more plausible literary hard sf into the writing curriculum promises to strengthen and complicate this integral connection between the popular cosmological imagination and astronomical science that may seem less readily apparent in the campy future visions of special-effects-dominated and action-oriented space-opera film and television. The recent renaissance of hard sf in space opera, for instance, pairs well with exploring the various Zooniverse offshoots and descendants of the original Galaxy Zoo citizen-science project. The discovery by eagle-eyed participants that some spiral galaxies do not have central bulges and yet still possess supermassive black holes at their center could be complemented, for instance, by the astrophysicist Gregory Benford's *Furious Gulf* (1994), a novel in his Galactic Center series of hard-sf space operas that grippingly and plausibly explore galactic black holes.

Likewise, the Zooniverse project Solar Storm Watch that asks participants to trace the edges of coronal mass ejections that could damage Earth satellites couples well with the short stories of Stephen Baxter's *Vacuum Diagrams* (1997), an installment in his Xeelee sequence of hard-sf space operas, and its riveting and rigorous depiction of sun-dwelling Photino Birds. Similarly, the project Galaxy Zoo Mergers that studies the collision of nearby galaxies might spark in a student a spirit of science-oriented inquiry, reflection, and keen insight that finds equal inspiration in Alastair Reynolds's series that begins with *Revelation Space* (2000) and the underlying cosmological answer to Fermi's Paradox—in a sentence, "Where is everybody?"—that extrapolates on galactic mergers overseen in the novels by so-called "Inhibitor" machine aggregates, termed by humanity, in a sinister fairy-tale way, "the Wolves," who actually wish to preserve life due to the eventual galactic collision of Andromeda and the Milky Way galaxies, pruning dangerously expansive technological progress. Yet even students less interested in exploring the intricacies of these cosmological extrapolations may nevertheless be sufficiently galvanized to conduct a deep, critical interrogation of issues related to citizen science and the military-industrial complex that has built up around space science in recent history.

One such lightning rod of boisterous student interest and excitement, especially for business and engineering majors, is a signal legacy of the Cold War space race, namely, the recent boom in space tourism and commercial spaceflight launched by Richard Branson's Virgin Galactic, Elon Musk's SpaceX, and Jeff Bezos's Blue Origin, among a wide array of small companies. Agonizingly slow in their gradual development of viable space travel, let alone their ultimate aspirations of moon bases and colonies on Mars, these companies contend daily with the overwhelming pragmatic difficulties of manned spaceflight. For instance, achieving the escape velocity necessary for launching a rocket into space, as well as making atmospheric re-entry if the rocket is efficiently designed to be reusable, presents extreme engineering hurdles for profit-oriented commercial companies. These issues are only compounded by the physiological dangers that the athletic human body must endure to become an astronaut, such as high-acceleration gravity stress, radiation poisoning, bone-density loss, muscle atrophy, a lack of temperature control, immunodeficiency, malnutrition, and stoppages of respiratory and cardiovascular flow.

As typified by a classic short story such as Cordwainer Smith's "Scanners Live in Vain" (1950), space opera has continuously grappled with the astronautical truism that "space is hard," a truism that the onset of space-tourism research and development has thrown into high relief. As explained by popular science journalist Rod Pyle, "to reach the higher hanging fruit [of commercial spaceflight] will require learning how to live in space in the long term, to mediate the effects of low gravity and radiation" (23). Likewise, due in part to the challenges and setbacks posed by the higher hanging fruit of commercial spaceflight ventures, hard-sf space opera has begun likewise to more heavily stress the extreme physical distress and menace of sending human bodies into space. This trend in space opera is vividly dramatized by the television show *The Expanse*, based on the novels by James S. A. Corey, the nom de plume of writing team Daniel Abraham and Ty Franck, and developed for television by the showrunner Naren Shankar, who holds a PhD in engineering and applied physics from Cornell University.

In the sixth episode of the second season of *The Expanse*, for instance, in an episode entitled "Paradigm Shift" (2017), based on the short story "Drive" (2012) by James S. A. Corey, the backstory of Solomon Epstein (Sam Huntington) appears. Approximately 137 years prior to the twenty-fourth century events depicted in the show proper, Epstein invents the fusion-reaction drive for starships, which permits the expansion of the human population into the Asteroid Belt, and the creation of the so-called Outer Planets Alliance (OPA), a radical working-class union who resist the hegemony of Mars and Earth. Epstein himself dies due to thirty-seven straight hours of prolonged acceleration and an end velocity of 5 percent of the speed of light during his accidental invention of the Epstein drive; and approximately a century and a half later, so-called Belters, physically thinner, taller, and with larger craniums due to living in low-gravity environments, can withstand prolonged exposure to High-Gs but experience extreme torture when subjected to the regular gravity of the inner planets.

The paradigm shift of the Epstein drive mirrors the discovery of the ancient alien "protomolecule" that drives the primary plot of *The Expanse* novels and television show, suggesting a direct linkage between the current entrepreneurial scramble for commercial spaceflight and the more exaggerated space-operatic speculations and economic-materialist realpolitik that shape the core of the epic narrative. In a voice-over montage that ends the episode, Epstein equates the radical breakthrough of his scientific novum and its predictive impact on the political-historical shape of the future history with a somber meditation on mortality itself: "Sooner or later, it happens to us all. Me, you, everyone we love. Maybe you see it coming. Maybe it surprises you." Grounded in the brutally hard engineering of mundane spaceflight, such a gritty, noir reimagining of familiar space-opera trope of interplanetary travel may furnish a bounteous trove of teachable ideas for student engagement with contemporary space culture.

AFRICAN CULTURAL ASTRONOMY, WAKANDA STRUGGLE FOREVER, P-FUNK, AND BINTI'S BILDUNG

De Witt Douglas Kilgore asserts that the 1950s American dream of spaceflight must be interrogated not merely as a maudlin dream of pluralism, tolerance, and rationalism projected onto the undiscovered recesses of the cosmos, what cultural critics have long lambasted as a naively utopian vision of exceptionalist American uplift and progress, but, more insidiously, as a thinly veiled function of systemic, historical white supremacist entitlement, hierarchy, and privilege. The contemporary salvaging of the dissident sf aspirational cosmological visions of minority cultures therefore holds the potential to reveal the ways in which an ethnonationalist ideology of "a freedom-loving people threatened by ubiquitous enemies is turned into a tale of a frightened, petty, tyrannical majority that has had the great fortune of encountering successive waves of obstinate minorities who dare to claim its promise of liberty and justice for all" (28). This trenchant critique of the space race revives the powerful cultural perceptions of disillusioned African Americans at the time of the Apollo program. As Mathew D. Tribbe writes, "The juxtaposition of the extravagant space program with the everyday deprivations of many inner-city African Americans showcased uncomfortable truths about the persistence of racism in the United States" (39).

In his recent rethinking of Marvel's first black superhero, the Black Panther, Ta-Nehisi Coates draws deeply on this countercultural vein of incisive protest that he first fully articulated in his award-winning memoir, *Between the World and Me* (2015), written in response to the most recent surge of police shootings of unarmed black people that triggered the latest incarnation of the Black Lives Matter movement. In this memoir, Coates writes, "That was a moment, a joyous moment, beyond the Dream—a moment . . . [that] originates in a view of the American galaxy taken from a dark and essential planet . . . the dungeon-side view of Monticello" (148). This lyrical vision, replete with nested allusions both to plantationocrat Thomas Jefferson's palatial estate and the hip-hop pioneers Public Enemy's album *Fear of a Black Planet* (1990), champions the blistering power of cultural dissent to puncture the banality of blandly progressive American rhetoric, while also commandeering the cosmological imagination to express a more deeply restorative commitment to the struggles of a black diaspora in dire peril.

In defiance of the obliterating traumas of slavery, the immensely popular discourse that is often labeled "Afrofuturism" has proliferated with, as Ytasha Womack explains, the imperative "to unearth the missing history of people of African descent and their roles in science, technology, and science fiction" (17). It is tempting therefore to apply this term Afrofuturism to an affiliated impulse in space science often referred to as an African-centered "cultural astronomy," or "archaeoastronomy," a vital research domain of modern astronomy dedicated to the cosmology, star lore, monuments, divination, data contributions, and everyday sky-gazing practices and understanding of the variegated ethnic groups indigenous to the diverse continent of Africa. The professional astronomers Johnson O. Urama and Jarita Holbrook describe their call for an African Cultural Astronomy Project as attempting to bridge the gap between scientific worldviews and the "rich astronomy" of indigenous African "architecture, folklore, myths, religion, [and] calendar" (50). Urama and Holbrook view this project as not only compensating for the currently marginalized cosmological beliefs and practices of African groups, foregrounding the regrettable invisibility of the underlying scientific systems of traditional African culture. This cultural-astronomy project specifically requests help for the fund-raising and construction of new astronomical research facilities in Africa, tangibly promoting the pursuit of astronomy to counter the currently neglected state of the discipline in African schools. If such citizen science were to achieve widespread success, ancestor worship, communal kinship ties, and the spiritual cosmology of indigenous African culture could be more readily framed alongside contemporary astronomical research as expressing comparable worldviews, thereby overcoming the unfortunate automatic conflation of space culture with white American exceptionalism.

One signal example of such African cultural astronomy that can illuminate students' contemplation of the cosmos comes from the scholar Aimé Dafon Sègla who demonstrates how the Ifa divination rituals and intricate cosmological traditions in the West-African Benin Republic imply a system of binary and hexadecimal numerical code. These mathematically precise astronomical rituals also intimate a political philosophy elaborating the Yoruba "euphemism of conquest" in which tribal power is allotted according to the prerogatives of a sagacious and shamanic leader who advocates a "tacit distributing and sharing role" (203) in public affairs and does not arbitrarily wield despotic military might and terror. Similarly, deeply investing in an Afrofuturist or black fantastic mode of writing—in which a traditional cosmology of spirit avatars, astral planes, and gods coexist with sophisticated space-age science and technology—Coates's *Black Panther: The Intergalactic Empire of Wakanda* (2018–2020) story arc seriously wrestles with fundamental problematics of power, freedom, race, and heroism set against a richly imaginative cosmic backdrop. A dynamic teaching text, the comic book complements the astronomical knowledge dominant in the global North with the neglected perspectives of African cultural astronomy that can be accentuated by a close, guided reading of Coates's sequential storytelling and artistry. Such textbook selection and curriculum promise to furnish students with a vibrant Afrofuturist critical lens through which to reconsider the insights and blindness of contemporary astronomical science culture at large.

Coates's first story arc *Black Panther: A Nation under Our Feet* (2016–2018) sets up its sequel when T'Challa, the Black Panther, or, in the West African language of Hausa, the "Damisa-Sarki," and hereditary king of Wakanda, endeavors to pacify a rebellion of disgruntled citizens that has arisen in the country, a fantastic African nation that has catapulted into far-future technological progress and simultaneously remained unconquered by the colonial scramble due to its resource cache of the rare, meteorite-derived Vibranium, a fictional sf mineral that essentially functions like a magical phlebotinum in the Black Panther mythos. To subdue the uprising, T'Challa seeks to forge a fraught alliance with the philosophy professor Changamire who advocates that Wakanda be transformed into a democratic republic as opposed to be ruled as an enlightened monarchy. Despite a surfeit of qualms and reservations, T'Challa succeeds in enlisting Changamire in publicly condemning a recklessly violent uprising inspired by Changamire's own radical teachings.

Yet this support cannot be secured without the professor first impressing on the king a copy of Edward S. Morgan's *American Slavery, American Freedom* (1975), an actual trailblazing work of popular historiography that famously argues "republican freedom came to be supported, at least in large part, by its opposite, slavery" (x). As stylized by the artist Brian Stelfreeze in a wordless, borderless panel at the bottom of the page of an up-angled extreme close-up beneath and behind Changamire's down-turned brooding face, T'Challa gets the professor to implicitly admit that the king heroically "draws no real joy from domination," unlike Changamire's ruthless disciple and rabble-rousing demagogue Tetu. While Changamire subtly insinuates that such naive heroic idealism makes T'Challa a flawed, conflicted, and even ineffectual monarch who neglects his duties at home while gallivanting around the world performing adventurous feats with the Avengers, the thoughtful philosopher still makes an impassioned speech on behalf of the beleaguered king, which proves vital in turning the tide of the self-destructive civil war erupting in Wakanda.

Inspired by Changamire's call for a turn to the democratic ideal of republican reform, T'Challa, for his part, also becomes overeager to emulate the American dream of infinite cosmic freedom and individualism on an astronomical scale, while overlooking the inescapable dependence of such an exaggerated libertarian ideology on the covert perpetuation of slavery, injustice, and genocide. *The Intergalactic Empire of Wakanda* begins with the amnesiac T'Challa waking up from a blow he receives resisting the laser prods of guards as a slave in the vibranium mines of the Benhazin star system, named for the King of Dahomey who fought an anticolonial campaign against the French at the end of the nineteenth century. T'Challa joins the resistance, called the Maroons, named after the militant escaped slave freedom fighters of the Black Atlantic, who are rebelling against the expanding intergalactic hegemony of Wakanda, led by Emperor N'Jadaka, that has ruthlessly conquered and enslaved several alien races, including the Kronan, Rigellians, and the Shadow People. Eventually, on the secret rebel planet Agwé, with the help of the sorceress Zenzi, who has become the avatar of the panther goddess Bast, T'Challa restores his memories by harnessing the power of an aquatic Lovecraftian creature called the Jengu, named after the water spirit of the Duala or Sawa ethnic groups of Cameroon. In this intricate mosaic of Afrofuturist storytelling, Coates deploys water imagery as a conduit for pan-African cultural memory, with its unavoidable historical associations of the Middle Passage, invoking as well the African cultural astronomy of water deities as a nexus of the material and symbolic realms, sky and earth, sacred and profane.

With his memories restored, illustrated in a marvelous splash page drawn by Kev Walker, consisting of a vortex of green lightning swirling out of T'Challa's forehead, and complemented by a collage of major scenes from the characters' life spanning over fifty years of canonical Marvel continuity, including an image from his first appearance of the character in *Fantastic Four* #52 (1966). This homage is significant because, as Adilifu Nama argues on the cultural role of the first major black superhero more generally, "Black Panther is steeped in the type of inventiveness and science prowess that rivals the genius of Reed Richards, a.k.a. Mr. Fantastic" (43). Such scientific ingenuity and resourcefulness subverts the invidious stereotype threat hovering over the primitiveness implied by the panther name, costume, and origin invented by two well-meaning white Jewish New Yorkers, Jack Kirby and Stan Lee, for a sensationalized, blaxploitation-era comic-book market that nevertheless economically profited off depictions of black people as "wild, bestial, hypersexual spectacles" (Brown 140). In *The Intergalactic Empire of* Wakanda, via the watery aid of the Jengu, T'Challa recalls that his brilliant scientist mother, N'Yami, was obsessed with astronomy, nursing "a dream to go journeying among them" in part out of a desire to discover the galactic source of Vibranium and protect the nation's precious resource from falling into the hands of Wakanda's enemies. Indeed, Coates stresses the overlooked scientific nature of black genius as well as the cosmological dimensions of the Afrofuturist imagination through an informational addendum to the comic that includes an astronomically precise map constructed by Coates and text that reads Agwé is a "Jovian planet whose mass consists mostly of liquid."

The artist Walker beautifully renders N'Yami's astronomical fascination with a panel composition showing T'Challa, traveling through time and space in the spirit realm called the Djalia, foreshortened and looking from a distance at his mother rapt by a computer screen and its mathematical three-dimensional representation of a star system that visually recalls Coates's addendum. T'Challa resurrects his mother's ambition, creating a detachment of Wakandans to explore the cosmos, and, unbeknownst to T'Challa, traveling back in time through a wormhole to set up a settler colony in the Benhazin star system. The forging of this empire originally occurs merely out of a mission of Wakandan self-preservation in hostile foreign space. Continuing a thread that weaves throughout Coates's run on *Black Panther*, T'Challa considers himself first and foremost a sophisticated scientist, not a political leader, and therefore decides to mount a space expedition himself to investigate what happened to his initial alpha flight only to discover that the Wakandan expedition has bloomed over two millennia into a fabulous, sprawling, draconian governmentality. Later in series, in conversation with T'Challa's ex-wife and lover, Ororo, better known by her moniker Storm from the *X-Men* comic series, T'Challa acknowledges his original desire for exploring the cosmos was devised in a "vain chase of [his] mother's shadow": "I regaled the dream. I indulged it. Until I was expelled from it and forced to see the truth"; Ororo tactfully agrees that it was T'Challa's lack of firsthand experience of slavery as "the underbelly of empire" that led to his latest misstep, a brutal reality that the mutant African goddess herself only discovered after visiting the United States as a member of the X-Men for the first time.

Not confined merely to visual or literary culture, the cultural astronomy of African cosmology has frequently been paired with the struggles of African American experience in popular culture at large. In the 1970s, for instance, the outrageous funk band Parliament-Funkadelic proved massively influential on popular music, especially on contemporary hip-hop soundscapes and fashion, by adapting the Afrofuturist legacy pioneered by the likes of Sun-Ra's Arkestra and the Jimi Hendrix Experience. Parliament-Funkadelic cannily tapped into the widespread hunger for futuristic talk of cosmic vibes and space odysseys for a black audience otherwise suspicious of what Norman Mailer famously diagnosed as the triumphalist White Anglo-Saxon Protestant ethnonationalism of NASA culture. Rickey Vincent suggests that "P-Funk's fantastic science fiction created a series of spectacular 'otherworlds' that Africans could inhabit freely . . . without fear of cosmic retribution, and whites simply did not exist" (312). The classic genre of funk music, from James Brown to Sly and the Family Stone, always had an association with the visceral blood, sweat, and tears of black bodies that the purity-driven white supremacist rhetoric of the silent majority feared and fantasized about as malodorous, unhygienic, abject, and unseemly, that is, in a word, "funky."

Yet, more recuperatively for its avid aficionados, funk represents the possibility of cathartic bodily release from the travails of being a black citizen of the world. Parliament-Funkadelic reworks funk from racist stigma through an exuberant fusion of the predominantly white psychedelic imagination of the time period with the music-genre protocols of funk, which sometimes intersected with Afrofuturist aesthetics. In the song "P-Funk" on the album *Mothership Connection,* for example, George Clinton ironically complements his colleagues in white-dominated psychedelic rock, before noting their conspicuous lack of the fastest and grooviest syncopation, promoting instead his fictional radio-astronomy station "We Funk" as the black communal solidarity of his funkier brand of trippy music: "Hey, I was digging on y'all's funk for a while/ Sounds like it got a three on it though, to me/ Like Doobie Brothers, Blue Magic, David Bowie/ It was cool, but can you imagine Doobie in your funk? Ho, we funk, we funk."

George Clinton would couple the Funkadelic sound explicitly to African cosmology by way of allusions to the astronomical functions of ancient Egyptian monuments and the popular urban legend that ancient aliens spawned the architectural splendor of Egyptian civilization. In the "Prelude" to *The Clones of Dr. Funkenstein,* Clinton vamps that cosmic funk has waited rediscovery in excavated pyramids "along with its co-inhabitants of kings and pharaohs." Clinton draws on the 1970s-era Egyptology craze to emphasize here the African roots of funk polyrhythm and semitonality, seeking a deep spiritual communion with the improvisational call and response and frenetic time changes of the African cosmology. The funk band indulges in a reverent pleasure for the leaky, excessive excretions and pulsating gyrations of the ritualized black body in funky motion on the dance floor. In *Mothership Connection* and its space-operatic follow-ups, the ancient African musical infections highlighted in Clinton's deep baritone vocals receives technologically adept updates from the "funksmanship" of Bootsy Collins's electric bass-guitar rhythms, Bernie Worrell's electronic-synthesizer throbs, Jerome "Big Foot" Brailey's pounding percussions, and Fred Wesley's densely orchestrated horn, saxophone, and string accompaniments. Parliament-Funkadelic musically champions the cosmically inflected richness and resilience of the black diaspora despite everyday onslaughts of adversity, oppression, and struggle faced outside the ecstatic sanctum of the dance floor.

Kudwo Eshun argues that the sophisticated mixing of tape tracks that Parliament-Funkadelic innovated also dovetails with the escapist camp aesthetics of the band: "The Parliamentary universe is frequently chided for its systematic silliness, its blatant impossibilities, its elaborately preposterous foolishness—so far removed from the reality of crime figures and prison statistics" (Eshun 145). The cryptic, fantastic cosmology of Parliament-Funkadelic's mythology includes UFO abduction narratives, references to extraterrestrial radio astronomy, tropes of evolutionary-biological uplift, and assorted technobabble of clones, ray guns, hallucinogens, alien romances, and interplanetary interchange and espionage, all elaborately encoding a grand world-saving imperative of rescuing authentic black experience in a dystopian poor urban environment. Parliament-Funkadelic continually hails its listeners to unlock its baroque Afrofuturist mythos, such that, as Jason Heller observes, "*Mothership Connection* scientifically enhance(s) funk so that it might cope with the rigors of an increasingly grim future" (111). The intricacy of the transcendent sonic grooves of Parliament-Funkadelic, its cosmic slop from the ancestral mothership of Africa itself, was matched only by the extravagant fashion of the band, platform boots, silver lamé jumpsuits, floppy hats and all, topped by its impressive stage pageantry in concerts, including a baby mothership that would fly over audiences before landing on stage and depositing the chief Afronaut himself, George Clinton.

Evident even in the pop excesses of Afrofuturism, a calm, abiding mindfulness about cultural astronomy promises to stimulate a critical, reflective understanding of the problems and paradoxes inherent to the cosmological imagination. Such a meditation on global African culture rejects the premise of astronomical research being oversold as purely a black box—that is, merely a congratulatory litany of discoveries consummated by Anglo-European civilization. Such a mindful balancing of both scientific and cultural perspectives remains more in keeping with educational trends more generally, since, as Ann Webster-Wright asserts, prioritizing a meditative pose in the classroom promotes a pedagogical stance in which "students are encouraged to probe assumptions, challenge arguments, critique certainties and analyze choices," (557) even those related to cherished scientistic pieties. Indeed, some literary sf explicitly addresses the reflective utility of such mindfulness pedagogy for understanding the cosmos from a perspective informed by cultural astronomy. Nnedi Okorafor's Hugo Award-winning *Binti* (2015–2017) series of novellas, for example, fuses the Afrofuturist imagination with contemporary scientific discourse that Coates and Parliament-Funkadelic also separately reconstruct. Okorafor, a Nigerian-American, however, has been outspoken in preferring the term "AfricanFuturism" to "Afrofuturism" since, as Coates also vividly illustrates, the latter term has more deeply woven ties to American history, even as it reaches out to diasporic influences, whereas Okorafor strives for representing an intense concentration on the contemporary culture of the African continent itself.

In the first instalment of the series, Okorafor stresses the parabolic arc of a young woman learning how to combine her scientific understandings of the universe with a profound cultural imagination from the very first scene forward. In this opening scene, readers are introduced to a young African girl, Binti, who prays for her technologically sophisticated transporter to rise from the sand (Okorafor 9) to steal her quietly away as she absconds to Oomza University, a space academy, in the dead of night. Her people, the Himba, hail from an African region in the present-day Namibia, and they cling to their vanishing cultural identity as being resistant to the technoscientific sway of the imperious galactic civilization at large. The Himba therefore roundly denounce Binti's decision to attend the university, given that the reverence for their local customs is so dear to them that they refuse to ever leave their homeland at all. The Himba even showcase such strict regional belonging on their bodily accoutrements: the Himba wear anklets to prevent snakebites and cover their bodies head to toe in Otjize, a mixture of butter paste and red ochre endemic to their native topsoil. Outside of Okorafor's fictional representation, applying Otjize to the skin and hair indeed constitutes a tribal practice of the Himba women of Namibia's Northwest region, protecting the skin from mosquitos, microbes, and sun damage as well as constituting a distinctive culturally African symbol of aesthetic beauty.

At Oomza Uni, a study group leader, and a member of the dominant Khoush group, regards Binti as an exotically "rare bizarre butterfly" (21) for her Otjize and anklets. Yet the group leader remains condescending and flummoxed over why Binti remains "covered in red greasy clay and weighed down by all those anklets" at the university, unconcerned with its cultural connection to the Himba "bloodline, culture, and history" (23). Relativistically harmonizing both science and culture, Okorafor reflects on and resists an uncomplicated coupling of the Himba with a romantic, neocolonial vision of the Himba as stigmatized markers of technological primitiveness or savagery; the Himba, after all, are fabulous technological savants of invention and mathematical mastery, especially renowned for their computerized astrolabes, inasmuch as the Oomza University starship is an organic synthesis of genetic engineering whose internal chambers are replete with a forest of oxygen-producing, toxin-absorbing plants. Moreover, when Binti communes with another techno-magical ancient Himba talisman, the Edan, she enters a state of mindful contemplation that is simultaneously alert and relaxed, embracing a receptive stillness or an intuitive Csíkszentmihályian flow of advanced scientific calculations: "Then my starved and thirsty brain dropped into a mathematical trance like a stone dropped into deep water" (38).

In Okorafor's series as a whole, Binti continually resists xenophobia, insularity, and invisibility. In the first novel, for instance, she brokers a truce between the Oozma Uni and the Meduse alien species by electing to be stung and hence transformed into the invading alien, her Otjize-covered hair converted to translucent blue tentacles called Okuoko. This alien mutation allows for a diplomatic rapprochement that reconciles the Meduse aliens and the human Earthlings, narrowly evading intergalactic war by ensuring the safe retrieval of the sacred stolen stinger that had been ignorantly exhibited in a museum by blundering Oozma Uni scholars. In the process of this cross-cultural coup, Binti loses her ancestral connection to her African homeland, depleting her last jar of Otjize and as a result at risk now of being ostracized as an outsider by her traditional Himba people. Through a combination of the ancient alien Edan device and the healing properties of her Otjize, Binti ultimately, though, saves the galaxy from an inconceivably destructive intergalactic war. Hence Binti straddles a dynamic cultural relativity that must continually adapt to changing cultural politics to survive. This fluid, heterogeneous subject position perpetually evolves in keeping with progressing scientific expertise yet remains ultimately faithful to the stigmatized cosmological worldview of her indigenous African heritage that the human experts at Oozma Uni dismiss as "tribal" (82). Through the de-escalation of hostilities originally triggered by the callous indifference of oblivious, culturally imperialist scientist-scholars, Binti blends astronomical scientific acumen and wherewithal with a contemplative mindfulness toward an alternative African-centered futurism and indigenous knowledge.

The second and third book in this easily assignable series, *Binti: Home* (2017) and *Binti: The Night Masquerade* (2017), likewise expands on the vexing dilemmas of cultural relativity, compromises, and diplomatic entanglements from the viewpoint of indigenous African characters who are simultaneously science-driven and tradition-infused. As Binti returns home, she must again negotiate the conflict between the Meduse and the neighboring indigenous Khoush group with the Himba people now caught in the collateral crossfire. In *Binti: Home,* Binti learns that her otherworldly mindfulness practice—that is, her mathematical trances—derives from "nanoids" (129) injected in the blood of her ancestors, the desert nomads known as the Enyi Zinariya, by benevolent alien visitors. Binti describes the calming trance as a cherished part of her Himba heritage, cementing her existential status as a "master harmonizer" whereby "you saw numbers and equations in everything, circulating around like eye floaters you see on the surface of your eye if you pay too close attention" (157). Likewise, in *Binti: Night Masquerade,* after having been killed endeavoring to broker a peace in a skirmish between the Himba, Khoush, and Meduse, and subsequently resurrected by melding with the gut bacteria of her sentient starship, New Fish, Binti, traveling through the magnetic resonances of Saturn's rings, atones with her African roots through the trancelike meditative process she calls "to tree," which she realizes is more than simply a literalization of an ancient Himba proverb: "A tree with strong roots laughs at storms" (130). The process of treeing with her edan is described as climbing higher into a vibrational astral plane of electrically humming harmony, where Binti discovers that her paradoxically alien African heritage traces its legacy back to "tall, humanoid gold people who loved the way the sun reacted with the Earth's atmosphere" (168). Binti has become one with the mothership connection.

OTHER WORLDS, OTHER LIVES: *MASS EFFECT,* LE GUIN'S HAINISH CYCLE, DO-IT-YOURSELF COSMOLOGY, AND EXOPLANET HUNTING

In 1995, future Nobel laureates Michel Mayor and Didier Queloz—by implementing advanced spectroscopic analysis of gravitational wobbles and luminosity variability in starlight—inaugurated the contemporary era of rapid discoveries of thousands of extrasolar planets (popularly known as exoplanets, for short). These discoveries regularly make headlines and capture the imagination of popular media, even filtering into videogame culture. The space-operatic videogame series *Mass Effect,* for instance, extrapolates on real-world exoplanet science to randomize the creation of planets, allowing gamers to launch probes that scan and mine for mineral resources as well as register statistics on the planetary size, compositions, atmosphere, temperatures, orbital periods, and so on. In an interview with NASA educator Sara Mitchell, the executive producer of the videogame, Casey Hudson, attests that the visual dynamics of the game developers were artistically and scientifically inspired by recent advances in astronomy: "The underlying idea was that real-life scientific discoveries consistently create imagery and concepts that blow away previous notions of what is possible." In an article for *Scientific American,* physicist Sean Carroll likewise applauds *Mass Effect* for this research-based game design for educational purposes, speculating that "someone might hear [a scientific term] as part of a game and then hear it again in a more scientific context, and that might help them ultimately gain a better understanding of what it is" (Greenemeier). In the writing classroom, an evaluative or analytic assignment that encourages students to assess the scientific foundations of an exceedingly popular game series such as *Mass Effect* can serve as a powerful motivator. This assignment has great pedagogical value especially if it's couched as complicating the typical responses of game reviewers and review aggregators that claim to approximate the wisdom of crowds yet frequently outsource critical thinking about videogame narratives to false-consensus notions of what vaguely counts as standard successful entertainment.

Indeed, the game theorist McKenzie Wark argues that the arbitrary inhabiting of the topological "gamespace" of a virtual world, such as probing the randomly generated exoplanets in *Mass Effect*, or actually making planetfall in land rovers that can drive around the story space or surveying the vast planets in spacesuits, meaningfully creates a productive cognitive dissonance in the immersive experience for gamers who must negotiate "the tension between enriched description and the poverty of storyline" (73). However, in *Mass Effect*, this tension is significantly tempered by complex character customization, nested role-playing narrative branching, and choice-driven dialogue options that at their most sophisticated can match the intricate open-world gameplay, sumptuous animated graphics, as well as the delicate mastering of the hand-eye coordination and reflex-based challenges of first-person-shooter mechanics. For instance, in the first *Mass Effect* game, one interesting culmination of a mission story line consists of players deciding whether their customized viewpoint character, Commander Shepard, should kill the last living Rachni Queen, the leader of an ethnocentric, militaristically metastasizing space squid empire, and effectively commit genocide on an entire alien species to preemptively ward off future threats to galactic civilization. This player's decision meaningfully informs the larger narrative design and choice structure of the game since the space-opera background revolves around the human species having been inducted as a new upstart member of the so-called Citadel, headquarters to a hybrid civilization that includes a robust commonwealth of diverse alien species, including the Salarians, Quarians, Turians, Asari, and Krogans. Hence the difficult political-ethical decision of how exactly to coexist with a variegated spectrum of alien species constitutes the baseline thematic concern of the game series; and if the player decides to release the Rachni Queen, in *Mass Effect 2*, messages from the Rachni provide advanced warning of the coming of a new implacable galactic threat to the pluralistic Citadel coalition, a threat consisting of the post-Singularity civilization-culling entities known as the Reapers. The densely convolved narrative of interspecies diplomacy and total galactic war extends itself, then, as a logical outgrowth of the science-based exploration of exoplanets in other features of the gameplay, such as the nonnarrative resource management and immersive open-world game dynamics.

As a result of this boom in exoplanet research, the PlanetHunters.org website helps hundreds of thousands of citizen-scientist planet hunters and amateur astronomers to harness the astronomical data from NASA's Kepler's Space Telescope and the Transiting Exoplanet Survey Satellite (TESS). These planet hunters then identify the crossings on spectroscopic light curves, which records monthly fluctuations in a star's brightness, when intervening planets pass in front of stars from our earthly perspective. For instance, in 2012, amateur volunteers Kian Jek and Robert Gagliano were credited as coauthors on a journal in *Astrophysical Journal* due to their work discovering the exoplanet PH1 from the PlanetHunters.org. Remarkably, PH1-b, or Kepler 64-b, a gas giant, orbits the quadruple star system, which includes one set of close binary stars orbiting in turn around another set of distant binaries. The scientific paper that announced this discovery begins with the statement of inspirational debt and human interest, "for decades the science fiction community has imagined that planets can orbit binary stars, yet only recently have such systems actually been detected" (Kostov et al. 1).

Interestingly, as a stimulus for productive scientific argument, as reported in *The New Scientist*, Laurance Doyle, of the SETI Institute in Mountain View, California, speculates that a rocky, cold moon of PH1-b could be home to alien life; however, Stephen Kane and Natalie Hinkel, of California Institute of Technology, caution that the complex, uneven habitable zones of such a world would lead to extreme, rapid freezes and thaws. This research constitutes only the most recent culmination of centuries of scientific argument on binary stars from William Herschel's accidental discovery of the phenomenon when attempting to measure the parallax of stellar superpositions, in the early nineteenth century, to the contemporary discussions revolving around determining the exact habitable zones around such stars or calculating the chaotic gravitational forces of such systems. Tracing the byzantine vagaries of such historical scientific debates over binary star formation—quite common cosmic phenomena, as it turns out—reveal opportunities for composition students to listen to opposed viewpoints, foster respectful dispute, patient consensus-seeking, and the inclusive changing of minds invited by dynamic astronomical inquiry.

Ursula K. Le Guin's *Hainish Sequence* of novels, and the relatively plausible verisimilitude of their extrapolations on planetary science, offer fictional literary representations of the cross-pollination of science-driven argument and ideas that a mindful approach to the cosmological imagination encourages. Le Guin wrote the interconnected but stand-alone Hainish cycle of novels as a loosely conceived cosmic genealogy whereby a Terran parent species disseminates various offshoots of genetically modified posthumanity across the universe and in the process creates the

League of Worlds through a faster-than-light communication device called the ansible. Although often problematically labeled a writer of "soft-SF classics" (McGuirk 109), due to being deeply influenced by the so-called soft sciences such as psychology, linguistics, anthropology, and sociology, in her nonfiction and interviews Le Guin also testifies to an abiding interest in probing the larger imaginative and cognitive implications of what is commonly perceived as the hardest of sciences, such as physics and mathematics.

Le Guin's technical innovations of fantastic world-building and conjectural thought experiment often therefore have profound astronomical underpinnings, given, as she states in the essay "Do-It-Yourself Cosmology," a professional sf writer needs to know not only "the arguments concerning the existence of planets in a binary system, and the probable effect of a double primary on orbits, tides, seasons, and biological rhythms" (122) but even "the usefulness of Do-It-Yourself Cosmology as a device for teaching the general principles, mechanics, and history of the cosmos, the solar system, and the planet Earth" (123). Anticipating the rise of citizen-science projects such as Galaxy Zoo and Planet Hunters, Le Guin's "Do-It-Yourself Cosmology" in the *Hainish Sequence* variously imagines scientific collaborations and teamwork in progressive, participatory, science-driven ventures that nevertheless become prone to cultural misunderstandings, diplomatic dysfunction, and political clashes. Likewise, in *Seeing like a Rover* (2015), the STS scholar Janet Vertesi considers the cosmological imagination of hard-scientific institutions and enterprises to be constituted by producing a dynamic social order dedicated to "the work of scientific representation" (8)—that is, contemplating the disagreements and compromises that compose and construe the imagined understanding of the cosmos.

The first Hainish novel, *Rocannon's World* (1966), takes place on a planet, called Fomalhaut II, orbiting the binary stars Formalhaut A and B, around which, as discovered only recently, an exoplanet candidate does indeed orbit. Le Guin juxtaposes the narrative of the novel, referred to as fantastic "legend," with excerpts from League of Worlds emissary Gaverel Rocannon's readings from *The League Handbook for Galactic Area Eight*, referred to as "equally true" (25) scientific data, including yearly orbital periods (approximately 800 Earth days), obliquity of the ecliptic (leading to marked seasonal change), gravity (less than Earth, hence flight is less cumbersome), atmosphere (oxygen-rich), distance to the sun (in the Goldilocks zone and therefore primed for the flourishing of diverse lifeforms). The handbook also mentions that the companion binary star of Formalhaut B is "visible as a superbright star" (25), which, as the reader learns through narrative-based world-building and contextualizing clues, the five separate sentient species of the planet call a sacred and navigationally useful "Greatstar" (11). Moreover, taking readerly pains to digest the cultural and personal implications of hard cosmological data, a close, scientifically informed audience may piece together that the excess illumination of two hot, bright, and young binary dwarf stars is consequential for many aspects of the fictional world-building, such as when it is revealed that an underground troglodyte species, sheltering itself from such excess illumination, has been balkanized from the variegated unknown surface areas of the world.

During his peripatetic misadventures and harrowing escapes on this mysterious planet, Rocannon himself eventually dismisses the handbook as "not much help" (53), projecting "savagely" (88) his hostility toward the empirical myopia and cluelessness of the progressive League of Worlds itself onto the harmless inanimate object of the book. Yet despite this frustration with the interstellar League being a theoretical, detached observer of the universe from afar, Rocannon, after his League colleagues are killed in combat, is relentlessly driven to make it to his ansible, a faster-than-light communication device, to alert the League of the hegemonic incursions of Formalhaut II by the viciously expanding Faraday empire. This sense of being guided by a critically utopian mission conscripted to the cosmological imagination entails what Vertesi helpfully analyzes as the cultivation of a generous rapport among the otherwise fractious members of the scientific community who can be nevertheless galvanized by the overriding imperative of "pursuing the unexpected, arousing and satisfying curiosity, and appreciating the sublime" (42).

The unpredictable creative insight of the scientific community derives from a contemplative practice that privileges empirical learning and champions sustaining "attention even in the face of paradox and contradiction with profound equanimity" (Zajonc 27). In pedagogical theory, such an emphasis on mindfulness and respectful listening for discrepant compromises, common ground, solidarity, and a frank sharing of perspectives has been coupled with teaching argument in a composition classroom (Peary 64), specifically the "invitational" or "Rogerian" mode of argumentation that is based on a "nonheirarchical, nonjudgmental, and nonadversarial framework" (Foss and Griffin 5) of argument-driven inquiry. This notion of a mindful communication with the divergent perspective of another subject position is manifested in Le Guin's sf universe through "mindspeech," a form of intensely intimate telepathic communication and a power instilled in Rocannon by one of Fomalhaut II's indigenous cave dwellers for his heroic tribulations (103). In turn Rocannon regifts this sophisticated meditation-based custom to the pluralistic galactic empire that is the League of Worlds, despite his reservations about the progressive nature of that cultural entity.

Challenging the scientific convention to jealously guard scientific data that can ensure prestigious careers, grants, and publications, citizen science projects often depend on the open-sourced sharing of datasets to encourage and sustain public collaboration, buy-in, and interest. As previously mentioned, Planet Hunters, for instance, uses the Kepler and

TESS telescopes, and Galaxy Zoo relies on the SDSS, one of the world's biggest telescopes whose rapid-fire, wide-angle lens can glimpse immense swaths of space from the far frontiers of the locally known universe. The SDSS has contributed to countless astronomical discoveries, but perhaps one of the most magisterial revelations was the Sloan Great Wall, an enormous collection of galaxy superclusters discovered by Richard Gott and colleagues when attempting to create a scalar map of the universe from the SDSS data. The notion that the shared astronomical project of mapping the unimaginably vast universe can evolve newfound cosmic perspectives also features prominently in the second published Hainish novel, *Planet of Exile* (1966).

In this novel, Le Guin again exhibits a profound cosmic imagination, returning to complex astronomical thought-experiments and world-building. Le Guin thereby continues the palimpsestic narrative of the discovery-driven interplanetary coalition called in later novels and short stories the Ekumen—from her influential anthropologist father Alfred Kroeber's Greek-derived term for the diffused inhabitants of a common culture, the "oikoumene." *Planet of Exile* takes place on the fictionalized planet Werel, or Alterra, a double planet, and the third planet of a Gamma Draconis system whose giant star, also known as Eltanin in our actual historical astronomy, was instrumental in the early confirmation of the Copernican heliocentric model of the solar system. Given the extreme luminosity of Eltanin, Le Guin speculatively extends the habitable zone of the star system, describing Werel as completing one orbit every twenty-four thousand Earth days, which, due to the planet's tilt, gives it oppressive winters that last fifteen Earth years in one stretch. Crucially, Werel is also home to vast oceans that can serve as vital heat sinks, and which can regulate an atmosphere and support life, given that these oceans are themselves tidally locked with a moon that orbits it every four hundred days, another feature that also greatly increases an exoplanet's habitability.

The post-Hubble vastness of the cosmic perspective Le Guin imagines grants the dwindling numbers of exiled League colonists a decadent patience and forlorn serenity, ostensibly forgotten as they are by the Hainish empire itself distracted by its own civil wars and invasions. The generations of "farborn" colonists, increasingly sterile, stillborn, and ostracized on the extreme environment of this alien planet, base their operations in the last vestigial stronghold Landin, which is geographically designed around a meeting hall that contains murals of Werel's star system: "On the east a stylized sun surrounded by nine planets faced the west wall's pattern of seven planets in very long ellipses round their sun" (139). Along with the libraries in the Records Room, this map of astronomical knowledge prominently enshrined and accompanied by clocks set both to local and galactically distant time zones displays a cosmic imagination that is dismissed as witchcraft and sorcery by the suspicious indigenous inhabitants of Werel and even doubted by some of the disillusioned Hainish exiles themselves.

Likewise, roused out his disaffected stupor by the stirring of romantic interest in the indigenous woman Rolery, the viewpoint protagonist Jakob Agat, the de facto leader of the colonists, overcomes his erstwhile passive detachment and lofty, quietist worldview to warn Rolery's father, Wold, and his tribesman of Tevar, of the rampaging hoards called the Gaal, despite the Hainish dictum of "the Cultural Embargo" that prevents intervention in alien development. After Tevar is sacked regardless of Wold's advice to his community, using advanced medical knowledge to care for the wounded, Agat helps the Teverans barricade in the fortified Hainish compound, waiting for the imminent planetary long winter to halt the besieging enemy. Le Guin thereby underscores the dynamic relativity of perspectives that these interloping, marooned spacefarers must negotiate, even while maintaining an "iron adherence" (143) to their scientific enterprise envisioned in their moldering cosmic maps, clocks, and libraries. Le Guin describes the depth psychology of the Hainish culture on this biologically inhospitable planet as abandoning their "old dreams of dominion" and instead eking out an ascetic existence by embracing a more profound contemplative mindfulness that values "the simple over the elaborate, calm over strife, courage over success" (142). Such egalitarian civic values parallel the gradual supplanting of a distant, elite science culture, with its aura of superiority and aloofness, and its domineering stance frequently accused of indoctrinating and propagandizing a recalcitrant public, by a more locally involved, reflective, subversive, and constructive citizen science.

The subsequent Hainish novel, *City of Illusion* (1967), set a thousand years after *Planet of Exile*, involves a traveler from Werel, Agad Ramarren, who assumes the identity of Falk after he lapses into amnesia when visiting a postapocalyptic Earth that has been conquered by the delusion-propagating Shing empire. Precipitating his mind restoration, his long-lost star-faring companion Orry explains the forgotten astronomical history that is "miraculous though not unusual" of the planet Werel to the mind-wiped protagonist: "Giant tides of the planet's deep seas obeyed a giant moon that took four hundred days to wax and wane; the world was rife with earthquakes, volcanoes, plants that walked, animals that sang, men who spoke and built cities: a catalogue of wonders" (321). To a student steeped in recent exoplanet science, Le Guin's lyrical description, in a rigorously extrapolative fashion, anticipates many of the debates and disputes still circulating around what precise exoplanet features should be searched for in the quest to discover life in the universe, including whether volcanic outgassing, tectonic shifts, and massive moons contribute to a planet's habitability quotient or not (Seager 579).

Equally important for understanding the literary representation of the scientific imagination in this passage is the way Le Guin frames this scientific knowledge as inextricably bound up with the cosmic rise and fall of the Hainish galactic civilization. Ramarren, as Falk, can only begin his perilous quest to restore his mindscape, and ultimately return voyage across the stars to warn Werel of the Shing threat, once he has laboriously relearned the Hainish canonical records that have been preserved by the indigenous Earthling community of Parth. A shaman of Parth teaches Falk about the polysemous slippages and indeterminacy of cultural and linguistic discourse that he should guard against: the infiltration of the Shing produces "the downfall of the League, the scattered community of man [that] had mistrusted trust and used the spoken word" (241). David Higgins argues that "if the League exemplifies a weak cosmopolitan hegemony that engages in neoimperial dominion, the Shing are the League's own dark reflection" (343). This critical insight can be extended to the cosmological imagination as much as the cosmopolitan one; the Shing, after all, have weaponized the "mind-lie," enforcing their sophisticated technological and scientific dominion by imposing a hyperreal schizoid simulacrum of historical reality on their subjugated subjects.

Likewise, in *City of Illusion,* Le Guin intimates that cultivating in the reader—and by extension the novitiate student writer—not just a reverence for but an interrogation of scientific topics for their subtle rhetorical and social convolutions can challenge ideological bad-faith narratives and misinformed arguments prevalent in media and culture. Scholars who research "agnotology," or the cultural dissemination of disinformation, label this posttruth misuse of science to create elaborate Potemkin villages of falsified knowledge (in the vein of what the Shing have perfected in the novel), as opposed to simply not knowing a factual reality due to attention fatigue or lack of education, the "active construction" of falsehood. The active construction of disinformation contributes to mass ignorance through "a concerted effort on the part of legitimized sources (e.g., scientists, pundits) to deceive, and to manufacture and manipulate individuals' and collectives' understandings of particular matters" (Block 35).

Students who approach value-laden scientific inquiry with a disciplined attitude of poised attention, alert mindfulness, or the wakeful curiosity and empathy of professionals in training must necessarily confront this active construction of misleading or deception-driven disinformation at large. In an argument-based essay assignment that requires scientific literacy for its research ballast—such as, for example, synthesizing and analyzing opposing viewpoints on exoplanet hunting—can help otherwise suggestible students negotiate from primary sources the doubt-laden, problem-solving process of scientific inquiry that in effect resists such ideological campaigns. This mindfulness approach to pedagogical citizen science helps restore confidence and trust in scientific institutions frequently dismissed as abstruse, insular, or complicit in systemic social injustice. The struggle for legitimacy and credibility in the public understanding of science today demands embracing, not evading or disavowing, the interpenetration of diverse cultural, social, and rhetorical perspectives as much as the conflicting scientific viewpoints and theories themselves.

Set centuries after *City of Illusion,* Le Guin's most beloved masterpiece, *The Left Hand of Darkness* (1969), winner of both the Hugo and Nebula awards, takes place. By this point in the potted galactic history, the Ekumen have recovered from Shing dominion and evolved past their earlier status as a peacekeeping force, flourishing into a mystic vision of anarcho-socialist utopia consisting of a league of eighty-three worlds. The Hainish envoy Genly Ai attempts to make first contact and induct the planet Gethen into the pangalactic fold, based on the mutual advantage not merely of a trade deal and its promise of "material profit" but for a more profound intercultural project of cosmic symbiosis that will contribute to the expansion of human knowledge: "the augmentation of the complexity and intensity of the field of intelligent life and the greater glory of God. Curiosity. Adventure. Delight" (413).

More experimental than the previous Hainish novels, *The Left Hand of Darkness* consists of a kaleidoscopic collage not only of Genly Ai's archival first-person account but also legendary fables or "hearth tales," scientific reports on the transgender hermaphroditic biology of the Gethenians, and an appendix called "The Gethenian Calendar and Clock." This document notes that Gethen's rotational spin, orbit, and lunar cycles are similar but slightly different than Earth but the Gethenian timekeeping practices are radically estranged, including numbering years backward and forward from the base year of the present. The interposed field notes of the incognito planet investigator Ong Tot Oppong also speculate that the most dominant factor of Gethenian life is not their hermaphroditism—since logically they have no outside reference point to which they can compare their unique sexuality and gender—but the overwhelming influence of the extreme permanent coldness of their planetary climate and meteorology, which may have even eliminated war since "the weather is so relentless, so near the limit of human tolerability even to them with all their cold adaptations, that perhaps they use up their fighting spirit fighting the cold" (458). Indeed, the extreme frigidity of what is also called the planet Winter informs variegated aspects of Le Guin's fabulously inventive world-building, including their Eskimo-esque tunics and trousers, the sixty words in their language for snow, the profound kinship of the hearth-sharing love rituals, and the razor devices that remove frozen foam from beer.

Through perhaps a combination of timely fact-checking research and sheer synchronicity in astronomical concerns, Le Guin again here anticipates another debate swirling around exoplanet formation that could stimulate directed student writing or spark discussion-oriented curiosity on this complicated subject. Some core accretion models argue that for a planet to become habitable of a certain massive size, it must be far away from a star such that solid-frozen material can exist within an orbital snowline, which prevents its transformation into an inhospitable gas giant. However, the discovery of gas giants close to stars, such as 51 Pegasus b, challenges such a snowline hypothesis and the existing theories of planet formation (Mayor and Queloz 358), and so arguments on behalf of competing viewpoints in this astronomical debate demand attention. In an analogous fashion, it may behoove a classroom to explore the interpretative possibilities about the literary and humanistic representations of science opened up by Le Guin's novel also remain unsettled.

The sf scholar Eric Rabkin reasonably refers to Le Guin's world-building in *The Left Hand of Darkness* as "environmental determinism" (157); however, Fredric Jameson perversely argues quite the opposite, namely, that "the cold weather of the planet Winter must be understood, first and foremost, not so much as a rude environment, inhospitable to human life . . . [but as] the fantasy realization of some virtually total disengagement of the body from its environment or eco-system" (60). Jameson thereby equates this posthuman isolation of the solitary individual on the inimical Gethen to the sf literary project of utopian "world-reduction," a distorted function of the repressed political unconscious of consumer capitalist alienation and angst. For Jameson, this aggrandizement of subjective dimensions of the private sphere comes at the expense of not only physical-material embodiment but also public agitation, protest, and engagement.

Such a Jamesonian hermeneutics of suspicion would certainly be in keeping with Le Guin's own nuanced exploration of Gethenian diplomacy and mysticism in the novel, especially the concept of "shifgrethor." This practice consists, after all, in supple, face-saving maneuverings of political power, or the Handdarrata religious sect's emphasis on the mindful synthesis and nonbinary thinking, and its complex nurturing of a mystique of unlearning. Such unlearning exploits the strategic value of conceding to nonrational "darkness" in mindful communication, or conceding to a deep-seated, sublime ignorance in the face of the totalizing complexity of an issue. Regardless of one's interpretation of the novel, though—that is, whether the empirical cosmic environment shapes the subjectivity of the individual indirectly, as Jameson counterintuitively argues, or an environmental determinism adheres more directly, as Rabkin more straightforwardly contends—what remains consistent in most analysis of the scientific worldview in the novel is that the cosmic imagination matters significantly to Le Guin's sf extrapolations and speculations. The reader is invited to contemplate the prospect that the flamboyant grandeur of the cosmic scale of the universe swallows the individual's idiosyncrasy into a vanishingly insignificant speck of relative sacrificial unimportance in the vast astronomical mechanism. Such a hydraulic plumbing of the cosmological imagination and its intersection with the construction of scientific knowledge is returned to with vigor in the subsequently published Hainish novel, *The Dispossessed: An Ambiguous Utopia* (1974).

The dueling threads of alternating chapters in *The Dispossessed*—one narrative thread set in the past on the planet Anarres, the other in the present on its estranged double-planet Urras—recount the Einsteinian theoretical physicist Shevek growing increasingly disillusioned from both of the societies endemic to these planets orbiting the star Tau Ceti. Anarres is the closest the novel comes to a critical anarcho-socialist utopia, even though it is rife with imperfections and dysfunction. Urras, on the other hand, admirably flaunts it conspicuous consumption as teeming with utopian superabundance on purely technological and economic grounds; however, in social and political terms, Urras is more tantamount to a capitalist anti-utopia, or even outright totalitarian dystopia, than its ambiguously utopian double planet foil. Indeed, the opening pages of *The Dispossessed* introduces a figural motif of erecting and unbuilding walls that define the flawed societies of both Urras and Anarres. On one important interpretative level, these barriers to scientific and empirical knowledge and progress will frequently recur in the narrative. For these walls may allegorize posttruth agnotology—that is, the disinformation, conspiracy theories, actively acquired scientific illiteracy, and deception campaigns—that prevents the successful dissemination of scientific discovery, inquiry, and understanding to the greater public.

In the deep background of the world-building, Urras had attempted a mining operation on Anarres, but the colonists of Anarres rebelled and gained independence. Since this political revolution, there has been a tense hostile cold war between the two planets, with little cultural or scientific cross-pollination or interchange between the radically divergent worlds. In the first scene, our protagonist, Shevek, a natural-born resident of Anarres, approaches the Port of Anarres in his planned interplanetary trip to Urras. Shevek defiantly breaks the embargo on any cross-cultural travel or immigration flow between Urras and Anarres. A mob of demonstrators line up to throw rocks at him since he is considered a traitor by some Anarresti citizens. The novel then follows Shevek running into a series of bureaucratic roadblocks and cultural impediments in both narratives in his grand quest to formulate and then widely communicate his general field theory of temporal physics that will eventually lead to the creation of the fictive science of a faster-than-light ansible communication device, which, as we Le Guin readers know, will allow the acceleration and leapfrogging of civilizational progress that permits the development of the Hainish League of Worlds. *The Dispossessed* tracks the cultural and historical circumstances surrounding the fictive science of a faster-than-light simultaneity principle behind the ansible devices, which serves to usher the equally novel cognitive estrangement of the conceptual breakthrough that is the Ekumen coalition. This sf paradigm shift of the simultaneity principle even seems to have some relatively plausible grounding in a futuristic extrapolation on contemporaneous real-world quantum mechanics, specifically the physicist John Bell's scientific paper "On the Einstein-Podolsky-Rosenstein Paradox" (1964) and the "uncomfortable" (Adler 210) hidden-variable conjecture, which suggests that faster-than-light information transmission might be theoretically possible (given that the spin of photons can be anticorrelated due to the Copenhagen interpretation about the probabilistic measurement of quantum particles and the law of the conservation of momentum). Hence, in *The Dispossessed*, Le Guin offers some of her most explicit and thorough novelistic interrogation of the construction of scientific knowledge regimes in theoretical physics and astronomy, and its wider linkages to civic discourse, public communities, and social institutions, that she would write in her highly accomplished and influential career.

In terms of their scientific rhetoric, citizen-science projects like Galaxy Zoo and Planet Hunters promote what Nielsen calls the serendipitous "modularity" (52) of small contributions in science, whereby even if an individual might be temporarily stymied by a difficult problem or bottlenecked by limited institutional resources, another individual can pick up the inertial slack, break this barrier, and lend assistance toward the overall iterative, shared scientific goal. In stark contrast to such a dynamic system of productive collaboration, in the flashback narrative of *The Dispossessed*, the leading Anarresti physicist at the Northsetting Institute on Abbenay, Sabul, plagiarizes the protagonist Shevek's revolutionary scientific ideas and refuses to let him teach or publish the science of those with whom the established éminence grise disagrees. This imperious censorship nearly drives the psychologically unstable Shevek to heartbroken suicide, as Shevek gradually realizes that his vital and original contributions to science are being stifled and suppressed by Saul's suffocating orthodoxy. By dint of a crucial conversation with his friend Bedap, Shevek articulates to himself the previously neglected dysfunctions and fractures within the anarcho-utopian planet they call home, even though the progressive, science-oriented society does seem to be ultimately headed toward a mutually desirable egalitarian goal. Their utopian ideals might ostensibly be socially progressive, but there is still a creeping crypto-authoritarianism prevalent on Anarres, a subtle bureaucratic centralization, replete with secretive, entrenched interests and an insidious will to power. Bedap self-righteously complains that mob rule prevents maverick minds from thinking for themselves, their radical independence crushed by an internalized acquiescence to popular opinion. In other words, the civic utopia, as perceived by its most adversarial critics, betrays its own democratic and egalitarian principles, scapegoating, ostracizing, or expunging perceived otherness on a scientific as much as cultural level with an oppressive power. The self-quarantined anarchistic society of Anarres has calcified into a rigid resistance to change that the simultaneity principle of the ansible and the subsequent policy of a more broadly inclusive cultural openness promises to upend.

Introducing students to the sociological and rhetorical underpinnings of the scientific enterprise through a close reading of Le Guin's novel promises to yield dynamic engagement in a classroom environment. Specifically, students may interrogate their own relationship to science by discussing the ways *The Dispossessed* dramatically explores the pleasures and perils of what Thomas Kuhn, in *The Structure of Scientific Revolutions* (1962), famously calls "paradigms" (Kuhn 11), or institutional frameworks of normative scientific knowledge legitimated by particular historical circumstances and social contexts, which inevitably squash dissenting "anomalies" or outliers, until a critical mass of such deviation leads to the supplanting, or transformative shifting, of the previous paradigm by a significantly new one. When undisrupted by paradigm shifts, the linear vision of modular scientific progress and social collectivity can be epitomized in our contemporary world by vanguard paradigms exploited by citizen science. Such citizen-science paradigms harness a flexible diversity of broadly dispersed expertise and industry that can scale up a domino effect of teamwork, consensus-building, and collaboration, resisting the isolated detours, idiosyncrasies, and inefficient discrepancies of the individualistic, egocentric scientist. Yet, just as Kuhn's discussion of paradigm shifts also implies, the eventual structural critique of bureaucratic and capitalist groundswell of scientific institutional change in Le Guin's scientific-materialist utopia seems to suggest that the public face of science is in perpetual conflict with radical anomalous breakthroughs at the frontiers of scientific knowledge; in a classic critical utopian tension, the needs of the many must be delicately balanced against the needs of the few—that is, the drive of the anomalous science and the demands of the collective scientific and social paradigm are eternally at loggerheads.

As powerfully recognizable to students, Shevek embodies this dynamic tension in his role as a long-suffering scientific genius desirous of expanding scientific knowledge whose submissions to scientific consensus are nevertheless regrettably necessary as a lubricant for social cohesion; therefore, even at the end of the novel, Shevek remains to a certain extent an irredeemable intellectual malcontent and lone scientist renegade who will never manage to successfully integrate into the collective social organism on Anarres. By comparison, his childhood friend, Tirin, likewise writes a subversive play and then is outcast into a manual labor working post by the Production and Distribution Coordination (PDC), which deploy work crews around the planet until he is forced to admit himself into a mental health clinic. This PDC is a supercomputer-ruled technological novum that is built into the world-building background of the novel, allowing for a new system of distribution and organization analogous to how sophisticated crowdsourcing of globalized knowledge works in contemporary citizen science. As discussed in chapter 3, students are also likely to automatically register the dystopian Big Brother aspects of such a gargantuan artificial intelligence.

Shevek, for his part, reconciles to his own predisposition to existential loneliness, intellectual self-exile, and emotional alienation by finding a compatible romantic life partner in Takver. These two rebels nourish a simpatico that is able to develop into a united front against the collective squashing of their individuality. In terms of the cosmological imagination of the novel, Shevek expresses their likeminded willingness to be joined together for the rest of their lives in terms of a counterintuitive paradox of his nonlinear, cyclical temporal-physics breakthrough: "Loyalty, which asserts the continuity of past and future, binding time into a whole, is the root of human strength" (879). Einstein's notion of relativity suggests time itself is a singular manifold that can be manipulated by gravity, defying clockwork notions of the fixed and absolute nature of Newtonian space and time, and which, in our own era, has led not only to the geosynchronous timing of satellites but the recent fruitful intersection of exoplanet hunting and citizen science vis-à-vis sophisticated theories of dark matter and dark energy. Similarly, and fantastically, Shevek's fictive future science of faster-than-light simultaneity defies Einstein's understanding of the universe in which the speed of light remains an unsurpassable constant. Takver and Shevek also achieve their own relative intersubjective simultaneity despite being physically separated for many years in the narrative arcs of their own lives; in their fraught but stable marital bliss together, they are responsible for a balanced harmony between each partner's modular contribution to each of their own individual progressive journeys as well as participating in the enlightened achievement of their mutually reinforcing domestic union as a whole. Shevek's epiphany at the end of the novel is described as a tranquil mindfulness of such multiple perspectives, as inhabiting a present-oriented immediacy and simultaneously viewing all temporality as the conflation of the arrow and a circle, ends and means, being lost and found, the journey and return: "There is nothing but the present, this Urras, the rich, real, stable present, the moment now" (890).

Disgusted by his own celebrity status on Urras, Shevek joins the resistance movement and participates in demonstrations that are violently suppressed, after he gives a powerful speech about his love of his homeland from which he has become a voluntary exile. Finally, then, Le Guin paints the anomalous utopian alternatives to current entrenched scientific paradigms as an ideal type of citizen science in transformative action, when Shevek seeks asylum with the Terran embassy on Urras. In the ultimate open-source act of relinquishment, Shevek gives the equations for the ansible to the League of Worlds for free and returns to Anarres with a Hainish ambassador from an Earth devastated by a future crisis of climate change and nuclear holocaust. This Terran ambassador views Urras as a lush and opulent utopia, forming no opinion on Anarres, except being overawed by its social experiment with collectivity, because the Ekumen have not been granted access to Anarres given its self-quarantining. At the close of the novel, as a result of his anomalous paradigm shift, Shevek's ansible helps to give hope back to the Hainish people sunk in political despair and their own planetary degradations. The novel then ends with Shevek returning home to Anarres "empty-handed" of his groundbreaking mathematical equations, remembering the mystic philosopher of his Urrasti people, Odo, who celebrates the paradox that all mindful people should strive for dispossession of the specter of competitive prestige and achievement but still nourish as well a desire-driven imagination of and curiosity for the natural universe. In other words, the transformative restructuring of the relationship between collective, crowd-sourced anarcho-socialism of grassroots science can be balanced against the routine social codes of existing normative scientific paradigms. As embodied by Shevek's life history, paradigm shifts enable the vibrant expansion of new perspectives and new participatory and progressive networks, projects, and citizen actors otherwise traditionally delimited in inaccessible silos of expertise, bureaucratic inefficiency, or microspecialized development.

SETI@HOME, BALL LIGHTNING, DUSTERS, AND SCIENTIFIC RHETORIC IN THE ORTHOGONAL UNIVERSE

One major facet of cultivating contemplative mindfulness in students involves encouraging an attitude of nonjudgmental "acceptance" and "openness" (Brown and Ryan 244) as opposed to lapsing into our default modes of dissociation and absentmindedness. As argued in this book, such a contemplative meditative presence can become ingrained as a core science-literacy skill through classroom discussion and essay assignments that wrestle with the pitfalls and promises of fostering successful citizen-science projects. The SciStarter organizer Darlene Cavalier highlights that effective citizen-science projects achieve an admirable responsiveness and flexibility by closely attending to the vagaries and flux of the dynamic scientific process itself: "People volunteering their time do not want to simply feel like they are helpful, they want to actually be helpful" (35). For example, the popular citizen-science project SETI@home, which has attracted over 5.2 million participants over two decades, announced in 2020 that it would cease sending out data to participants. Starting in 1999, SETI@home, hosted by the UC Berkeley, uploaded a screensaver onto the home and work computers of legions of participants, which harnessed the processing power of these otherwise resting machines to crunch narrow-band radio data from the Arecibo Observatory in Puerto Rico to search for signs of extraterrestrial intelligence. SETI@home initiated successive waves of so-called BOINC projects that follow its model of harnessing the distributed computing of public volunteers who are willing to lend their underutilized computing resources to scientific endeavors.

In a press release posted to the SETI@home message board on March 2, 2020, this long-lasting public effort was put into "hibernation" because of "diminishing returns" before expressing optimism in "look[ing] forward to what comes next." If the currently beta-testing affiliated Astropulse project, which expands the band of radio signals analyzed, is any guide, the rebooted SETI@home project of the future will significantly broaden the search methodologies for unknown astrophysical phenomena beyond what Paul Davies, chair of a postdetection task group at the International Academy of Astronautics, in *Eerie Silence,* diagnoses as the anthropomorphism inherent in "SETI parochialism" (14). Davies argues for the widening of the parameters of SETI searches in the future to encompass the possibilities of alien physiologies radically divergent from human species, including increasing our capacity to detect an enlarged range of biosignatures—such as minerals, chemicals, microbes—and technosignatures such as radiation leakage from engineering megastructures.

Demonstrating such an adaptability in the SETI@home project would confirm the Chinese sf writer Cixin Liu's touching tribute to the noble aims of citizen science, and withering critique of botched citizen science, in his novel *Ball Lightning* (2005). In this novel, Chen, whose parents are killed in a freak ball-lightning occurrence, develops an all-consuming scientific obsession with investigating this strange atmospheric phenomenon. At one point in his lifelong quest to understand ball lightning, goaded on by the unscrupulous Lin Yun, Chen becomes inspired by the SETI@home precedent, and hesitantly buys into the dubious analogy that the search for ball lightning could be equated to the search for intelligent life, only to fail to carefully build a dedicated pool of eager volunteers for his "Search for Magnetic Life at Home" website. The deeply obsessed Chen then realizes "how weak moral constraints are when you crave something" (101), hacking into the SETI@home servers to siphon off their data-crunching resources. Exhilarated by the initial results, Chen is then quickly crestfallen, once his dishonesty is duly caught, and he is upbraided and chastened by the director of SETI@home for such "brazen intrusion" (104). In this concise parable of both the siren seductiveness of citizen science and its potential for the unethical misuse of volunteer efforts, Cixin Liu foreshadows Chen's later more devastating susceptibility to the femme-fatale Lin Yun's militaristic weaponizing of ball-lightning macro-fusion energy, one of Cixin Liu's trademark speculative-science flourishes. When assigned in a composition classroom, this novel therefore promises to percolate a fruitful and dynamic meditation on citizen science projects being both attentively responsive to and constitutive of the good-faith contributions of its actively engaged participants.

In 1999, NASA launched the *Stardust* spacecraft with the ultimately triumphant goal of capturing dust samples from the comet 81P/Wild 2. One pragmatic reason undergirding the seemingly pure, abstract scientific curiosity that drives such a mission is to understand how to prevent or mitigate the potentially catastrophic collision of Earth with a comet, asteroid, or meteorite, collectively grouped together as Near Earth Objects (NEOs); hence NASA has also initiated the Safeguard Survey, which has managed to track 90 percent of NEOs that are 140 meters in diameter—the Cretaceous-Tertiary-Extinction asteroid that most likely killed the dinosaurs was 15 kilometers in diameter. Viewed as especially urgent following the relatively minor injuries and damage that resulted from the Chelyabinsk meteorite in 2013, which consisted of a NEO totaling approximately 20 meters, the scientific probing of the exact material size, shape, and composition of large space objects may help to thwart the near certainty of a species-extinction-level collision in Earth's future, as scientists speculate on whether interventions such as "mirror bees," or laser ablation, might divert the orbital dynamics of such incoming threats. Once *Stardust* passed through the tail of the comet with its racket-shaped aerogel collectors and returned back to Earth safely, NASA turned to citizen science to recruit tutorial-trained volunteers, so-called "dusters," to observe and identify the hundreds of thousands of microscopically enhanced interstellar grains that measure in the miniscule microns in size picograms in weight.

Stardust@home, as the NASA citizen-science project was called, rewards its thousands of dedicated, vetted citizen volunteers with coauthor credit on scientific papers and the opportunity to name observed deep-space particles. Even more significantly, though, the findings of Stardust@home have triggered vigorous debates that fundamentally reframe how comets are scientifically understood. Surprisingly, through sophisticated analysis and painstaking scrutiny, it was eventually determined that the probe collected large rocky particles, called chondrules, and organic matter, such as the amino acid glycine, that needed to be formed closer to the sun than the traditional "dirty snowball" theory of comets allowed. Thus scientists were confronted with a dilemma of resolving "how high-temperature 'asteroidal' dust was transported from the inner Solar System . . . in a few million years over huge distances" (Starkey 165), ultimately leading to competing theories that comets might have been affected by turbulent jets or high-temperature patches of giant planetary embryos in the protoplanetary disc. Curating for curious students the scientific debates swirling around Stardust@home, while closely pairing such discourses with more rhetorically engaged (and elaborately entertaining) sf texts, promises to simultaneously encourage civic-mindedness and complicate the public understanding of science in the writing-driven classroom.

Some richly complex hard-sf novels for such a pairing are Greg Egan's *Orthogonal Trilogy*, consisting of *The Clockwork Rocket* (2011), *Eternal Flame* (2013), and *The Arrows of Time* (2014). The trilogy, after all, revolves around desperate efforts by a band of visionary scientists to save their species from their predicted planetary demise because of the looming impact of species-killing asteroids named the Hurtlers. On Greg Egan's website, moreover, this trilogy has its own built-in pedagogical apparatus constructed for readers of various scientific backgrounds, including headier forays for enthusiastic science and math majors, and gentle introductions to the conceptual underpinnings of the novel's exotically imagined environments for students with rusty or nonexistent backgrounds in physics and math. Generally, guiding students stepwise through the densely worked-out, credible science experiments and cosmological theories in these difficult novels is a robust antidote to the cursory and superficial narratives of sublime wonder and mystery over the vastness of the cosmos that may strike some students, after years of such academic and popular interpellations, as both maudlin and boring. Likewise, encouraging students to merely skim and scan certain sections of the novels that defiantly indulge in fairly recondite diagrams and abstract lectures may grant students more agency and self-determination when approaching the remote technicalities of scientific and mathematical subjects, while also delicately exposing them to this rigorous level of professional discourse. Egan's complexly realized alternative-physics thought-experiments can also be more philosophically approached by students, or instructors for that matter, who do not have the luxury or background for tackling the specialized arcana of math and physics in all its daunting intricacy. Philosophy students, for instance, may be enthusiastically interested in the general metaphysical problem of time-symmetry that inspires the trilogy, and thus they may want to consider arguments on the idea that "what appears in ordinary physics as an immutable law of nature (viz. the Second Law of Thermodynamics with a given arrow of time) . . . might not be true in all universes, even if the underlying fundamental physical laws are the same" (Ellis 1240).

Another fascinating dimension of the novels worth foregrounding in a composition classroom involves its modeling of contemplative citizen-science rhetoric—that is, the art of mindful listening to multiple, divergent viewpoints both to dispel misunderstanding and to provoke dialogue on the overlapping domains of scientific and public matters. As argued earlier in this chapter, to obviate the seeming dependence of citizen science on a reductive and dehistoricizing version of a scientific worldview, teaching attentive listening to the rhetorical features of science might be helpfully framed as being akin to meditation, specifically so-called "lovingkindness" meditation or, in more academic terms, "listening rhetoric." Such meditation practice opposes itself to "our pop-cultural ideas of love as mushy, related to wanting, owning, and possessing"(Salzberg 18), and instead entails the belief that concentrated meditation involves generously and actively learning from those with whom we disagree or from whom we differ.

Influenced by Kenneth Burke's critique of rote scientism, the foundational composition-studies scholar Wayne Booth, in *The Rhetoric of Rhetoric* (2004), memorably divided listening rhetoric into a useful schema, including "rhetorology," the aspirational searching for legitimate common grounds between reasonable disputants; "dogmatism," the unassailable commitment to passionate convictions; "rhetrickery," the pretext of listening for manipulative, deceptive ends; "self-censorship," the surrendering to an unreasonable perspective out of fear of retribution or violence; and ill-advised listening for which changing one's mind would only disastrously affect the listener. According to Booth's taxonomy, only rhetorology constitutes efficacious rhetorical attempts at meaningful accommodation, adjustment, and adaptation that transcend rhetrickery, dogmatism, self-censorship, and counterproductive listening, which are all variously tantamount to "waffling, catering, sucking-up, shoddy spinning, or plain unforgivable lying" (52).

Exhibiting an astonishingly consistent literary treatment of the democratizing rhetoric of citizen science in society, Greg Egan's project of science literacy, science-advocacy, and science-education through innovative sf literature vividly illustrates Booth's listening rhetoric, both of the deliberately bad-faith and more rhetorically productive variety. The first book in the series, *The Clockwork Rocket,* lucidly introduces readers to the cognitively estranged parameters of an impossible universe. As a result of this experimental hard-sf technique, Egan rigorously contemplates the physics, chemistry, and biology necessary for one of the most outworn and rote tropes of space opera, namely, the faster-than-light starship drive, which is often derided by savvy readers who have been correctly taught that that such travel is impermissible according to contemporary theoretical understanding of the speed of light as an unsurpassable constant. In the alternative cosmology of Egan's fictional universe, by contrast, contra Einstein's special relativity, time is not treated as usefully analogous to a "spacelike" dimension; rather, on an astronomic scale, time operates exactly equivalently to the three spatial dimensions in a mathematically precise fashion. This slight tweak in how the universe works results in a wide variety of marvelous and bizarre consequences, including variable speeds of light, the possibility of traveling at infinite velocity, time travel in which moving sufficiently faster makes time slow down rather than the opposite rule of a velocity-induced temporal speeding-up that our empirical universe observes, a universe shaped like a four-dimensional sphere, antimatter asteroids, plants emitting instead of absorbing light, and a sentient quadruple-eyed species so chemically unstable that they extrude limbs at will and reproduce by a self-annihilating fission into offspring.

The scientist-hero protagonist of *The Clockwork Rocket,* Yalda, is a female member of this species and therefore vulnerable to the terrifying childbirth-death as well as the vicious patriarchy and discriminatory sexism of the culture that has evolved from this biological condition. Both a once-in-a-generation genius of a scientist and a brilliant practitioner of rhetorology, Yalda eventually takes controversial contraceptive pills, while overcoming a series of dogmatic rhetors with whom she engages in high-level academic disputes for the purpose ultimately of constructing a generation starship, *The Peerless,* to investigate the Hurtler asteroid phenomenon and rescue her home planet from an imminent collision. For instance, when challenged by the powerful and respected Ludvico, an established masculinist defender of the so-called "corpuscle" theory of light, the upstart Yalda, a recent convert to the wave theory of light, dutifully accepts advice to display tact and diplomacy when arguing against her opponent, whereas Ludvico imperiously upbraids Yalda for waffling—that is, succumbing to her "hubris" and "inconstancy"—while for his part dogmatically adhering to his own increasingly beleaguered position, refusing to be "driven to the same undignified series of self-contradictory declarations and endless changes of allegiance" (Egan 51). Yalda's groundbreaking wave theory of light, however, eventually prevails as the scientific consensus for a generation, illustrating her deft, flexible resistance to dogmatic scientific rhetoric, as she reasonably aspires to accommodate the next logical extension of a cogent theory, mounting evidence, and convincing refutation.

Another striking example in *The Clockwork Rocket* of Yalda patiently and contemplatively listening to alternative points of view with which she disagrees occurs when she must adjudicate a dangerous act of engine sabotage on board *The Peerless.* Yalda's crew members urge the execution of the captured saboteur, Nino, a poor-farmer pawn in a larger dynastic-capitalist business feud. Yet Yalda believes rehabilitation is possible and cannot bury this impulse in a shroud of self-censorship, or sucking up, for the sake of preserving her stabilizing authority on the generation starship. As heroic as her dogged pursuit of scientific inquiry proves to be, Yalda also becomes a beacon of circumspect scientific rhetoric, refusing to allow the threat of violence or the extreme assertion of political power dictate her decisions, even when she has become the figurehead of scientific-rationalist emergency privilege and authority on the ship. Instead Yalda puts Nino in prison partly for his own safety, becomes his personal tutor, and smuggles in supplies to help him write down the mythological sagas of their common culture that he has remarkably memorized for their ship library. Still her sworn enemy, despite Yalda's righteous adjustments to understand his opposing perspective, Nino initially feels Yalda's benevolent protection of him is patronizing and hypocritical; nevertheless, during the slow progression of three years of imprisonment, Yalda builds trust with Nino, and a close relationship unexpectedly blossoms. When Nino is finally released from confinement in his cell, in part to conserve limited resources following a harvest blight and in part to die with dignity, Yalda and Nino consummate their intimacy and become "co-steads," ensuring Yalda's voluntary self-immolation, but not before Nino agrees to respectfully transmit her legacy to future generations. The novel ends with the lyrical description of a miraculous discovery of literal astrophysical common ground between two erstwhile enemies: "They were sharing light, and the light carried Nino's promise to protect what [Yalda] would become" (319).

Three generations after *Clockwork Rocket*, in *Eternal Flame*, a new paradigm-shifting scientist, Carla, confronts inertial resistance to conceptual breakthroughs. While discovering what in the Orthogonal universe is the equivalent of quantum mechanics—specifically, postulating the existence of "free luxagens," echoing Arthur Compton's early twentieth-century observation of the X-ray scattering of free electrons in graphite—Carla sees the "rhetrickery" that shores up the edifice of the textbook theory of energy dynamics begin to crumble. Carla therefore no longer assents to spin-doctoring the illogical contortions of the widely accepted theory in light of these empirical results, despite the revolutionary, counterintuitive implications of her own alternative quantum theories: "One anomaly was an embarrassment, two were perplexing . . . but a dozen or so might come together to reveal a whole new vision of the world" (43). The theoretical physicist Carla's discoveries, and their path to the eventual invention of a self-sustainable eternal flame of a novel fuel source, as well as a deeper understanding of the orthogonal matter of the extinction-threatening Hurtlers, parallels the radically disruptive discoveries of her biologist co-stead, Carlo, whose experiments manage to make it so women can give childbirth to a single child without dying. Yet despite abductions, forgeries, and other hardball political jockeying by extremist patriarchal factions opposed to granting a woman a right to choose, Carlo also refuses to budge in his scientific convictions, never acquiescing to the duplicitous opposition, given the urgency of the stakes heightened by a blight on the contraceptive crops, which makes women starve themselves so as to not limit their reproductive capacity.

In *The Arrows of Time*, Egan recuperates the mindfully pluralistic scientific process of hopeful inquiry and resilient argument from an overly predictive determinism wherein rhetorical accommodation, adjustment, and adaptation becomes all but impossible. Given the fixed nature of the temporal dimension in this universe, *The Peerless* creates an intricate time-traveling messaging system to communicate with the future versions of themselves, thus violating ethical principles of knowing one's preordained fate possibly infringing personal agency and volition. To decide whether to use this messaging system, the reigning democratic council conducts debate between Agata, who wishes to legalize time-traveling messaging, and Ramiro, who wishes to outlaw the practice. Ramiro argues that the messaging system would silence debate: "Knowing even the most mundane facts from the public record would crush our political lives, flattening our inner dialogues into a choice between impotent rage and apathetic conformity" (82).

Agata counters in favor of communication since it would galvanize the crew for their eventual return home, even warnings would not be debilitating since "most of which would be no more harmful to us than a friend's reminiscence about a youthful misadventure that we'd prefer to forget" (83). In a close vote, Ramiro fails to convince the council to ban the messaging system, but ever the "peacemakers, reaching out to their enemies for the sake of the greater good" (112), Ramiro is permitted to accompany a dissident search party exploring whether colonizing an orthogonal planet is advisable, especially for those opposed to the grandfathered-in status quo of uncles caring for children of deceased mothers. At the end of the book, following this psychedelic scientific expedition, Ramiro again struggles with this same predestination paradox when he supports the terrorist sabotaging of the time-traveling messaging system, since, confusingly, the recent absences of messaging from the future implies an active sabotage would at least mean that the ship was not destroyed by a Hurtler asteroid. This sabotage is narrowly prevented by Agata who not only rescues agency from a fatalist surrendering to foreknowledge but also champions rhetorical accommodation regardless of the looming threat of violent backlash and dangerous social instability. Indeed, Karen Burnham compellingly argues that the ending of the trilogy marks "the triumph of citizen action over both entrenched political interests and extremist violence."

The citizens of the Orthogonal universe essentially establish a pragmatic scientific community of dissident technological utopians who forge progressive affinity-based coalitions based on likeminded enlightened civility conjoined to rationalist-empirical rhetoric and logic. In *The Eternal Flame*, Egan neatly encapsulates the science-educational mission of his entire fictional project when a historian defends his vocation by stating "the most powerful engine in history" is "people arguing about science" (361). In popular culture, and in broadest outline only, Egan's rigorous extrapolations on cosmology and astrophysics are not therefore dissimilar to the campier, more exaggerated, and brashly space-operatic speculations of the pluralistic, science-driven Federation Starfleet from the *Star Trek* franchise. In the "Context Is for Kings" episode from the show *Star Trek: Discovery*, for instance, with humans at devastating war with Klingons, ship captain Gabriel Lorca (Jason Isaacs) defends the ethically questionable pursuit of a mycelium spore drive to the soberly hesitant Michael Burnham (Sonequa Martin-Green).

The notion of a spore drive is based on the speculative scientific work on fungi and dark matter theorized by the actual mycologist Paul Stamets, whose ideas are refracted into a fictionalized "astromycologist" Paul

Stamets (Anthony Rapp) in the television show. Over a stunning holographic array, and in the typical *Star Trek* educational gesture of an enlightened, utopian exploration of new worlds, new civilizations, Captain Lorca implores Michael to awaken a spirit of scientific discovery: "Imagine the possibilities. Want to see where they [the mycelium spores] are going? Where they've been? Or where they could take us?" That Lorca turns out to be a disguised warmongering doppelganger from a neo-imperial, technofascist mirror universe in the overarching narrative of the show only reinforces the dire need for popular civil discourse that constructively meditates on, deliberates over, and scrutinizes exploratory scientific progress as a lodestone of the popular imagination. Indeed, in the bright, not-too-distant future, it is conceivable that science as an enterprise may be popularly understood not merely as a pure and abstract methodology existing in some irrefutable political-cultural vacuum of space but instead as a meaningful conduit for significant personal engagement and public civic-mindfulness.

Works Cited

Adler, Charles. *Wizards, Aliens, and Starships: Physics and Math in Fantasy and Science Fiction*. Princeton University Press, 2014.

Anders, Charlie Jane. *The City in the Middle of the Night*. Tor, 2019.

Andre-Driussi, Michael. *Gene Wolfe's* The Book of the New Sun. Sirius Fiction, 2019.

Annihilation. Directed by Alex Garland, performances by Natalie Portman, Jennifer Jason Leigh, and Tessa Thompson. Paramount, 2018.

Aoun, Joseph. *Robot Proof: Higher Education in the Age of Artificial Intelligence*. MIT Press, 2017.

Asimov, Isaac. *Foundation's Edge*. Del Rey, 1982.

———. "The Myth of the Machine (1978)." *Asimov on Science Fiction*. Doubleday & Company, Inc., 1981.

———. "The Science-Fiction Breakthrough (1979)." *Asimov on Science Fiction*. Doubleday & Company, Inc., 1981.

Asimov, Isaac, and Frederik Pohl. *Our Angry Earth*. Tor, 1991.

Atwood, Margaret. *MaddAddam*. Anchor Books, 2013.

———. *Oryx and Crake*. Anchor Books, 2003.

———. *The Year of the Flood*. Anchor Books, 2009.

Aungst, Heide, Jennifer R. Fishman, and Michelle L. McGowan. "Participatory Genomic Research: Ethical Issues from the Bottom Up to the Top Down." *Annual Review of Genomics and Human Genetics*, vol. 18, 2017, pp. 357–67.

Badmington, Neil. *Alien Chic: Posthuman and the Other Within*. Routledge, 2004.

Ballard, J. G. "The Coming of the Unconscious." *A User's Guide to the Millennium*. Picador, 1996.

Bell, Kevin. *Game On! Gamification, Gameful Design, and the Rise of the Gamer Educator*. John Hopkins University Press, 2018.

Block, Cathy, and Gerald Duffy. "Researching on Teaching Composition: Where We've Been and Where We're Going." *Comprehension Instruction: Research-Based Best Practices*, edited by Cathy Collins Block and Sheri R. Parris, Guilford Press, 2008, pp. 19–37.

Block, David. *Post-Truth and Political Discourse*. Palgrave Macmillan, 2019.

Bonney, Rick, and Janis L. Dickinson. "Overview of Citizen Science." *Citizen Science: Public Participation in Environmental Research*. Cornell University Press, 2012.

Booth, Wayne C. *Critical Understanding: The Powers and Limits of Pluralism*. The University of Chicago Press, 1979.

———. *The Rhetoric of Rhetoric: The Quest for Effective Communication*. Blackwell Publishing, 2004.

Borges, Jorge Luis. *Collected Fictions*. Penguin Books, 1999.

Borski, Robert. *Solar Labyrinth: Exploring Gene Wolfe's* Book of the New Sun. iUniverse, 2004.

Bould, Mark, and Sherryl Vint. "Dead Penguins in Immigrant Pilchard Scandal: Telling Stories about 'the Environment' in Antarctica." *Kim Stanley Robinson Maps the Unimaginable: Critical Essays*, edited by William J. Burling, McFarland, 2009, pp. 257–73.

Briggs, John. "Writing without Reading." *Write It: A Process Guide to College Essays, with Readings*, edited by Linda Strahan, Kathleen Moore, and Michael Heumann, Kendall Hunt, 2017, pp. 616–25.

Broderick, Damien. *Psience Fiction: The Paranormal in Science Fiction Literature*. McFarland, 2018.
Brown, Jeffrey. "Panthers and Vixens: Black Superheroines, Sexuality, and Stereotypes in Contemporary Comic Books." *Black Comics: Politics of Race and Representation*, edited by Sheena C. Howard and Ronald L. Jackson, Bloomsbury Publishing, 2013.
Brown, Kirk Warren, and Richard M. Ryan. "Perils and Promise in Defining and Measuring Mindfulness: Observations from Experience." *Clinical Psychology*, vol. 11, no. 3, 242–48.
Buhrmester, Michael, Tracy Kwang, and Samuel D. Gosling. "Amazon's Mechanical Turk: A New Source of Inexpensive, Yet High-Quality, Data?" *Perspectives on Psychological Science*, vol. 6, no. 1, 2011, pp. 3–5.
Burke, Kenneth. "Four Master Tropes." *The Kenyon Review*, vol. 3, no. 4, Autumn 1941, pp. 421–38.
Burnham, Karen. "Free Will in a Closed Universe: Greg Egan's Orthogonal Trilogy." *New York Review of Science Fiction*, no. 307, March 2014, www.nyrsf.com.
Butler, Octavia. *Lilith's Brood*. Warner Books, 1989.
Canavan, Gerry. *Octavia Butler*. University of Illinois Press, 2016.
Carillo, Ellen C. *Securing a Place for Reading in Composition: The Importance of Teaching for* Transfer. Utah State University Press, 2014.
Carr, Nicholas. *The Glass Cage: How Our Computers Are Changing Us*. W. W. Norton & Company, 2014.
Cavalier, Darlene, Catherine Hoffman, and Caren Cooper. *The Field Guide to Citizen Science: How You Can Contribute to Scientific Research and Make a Difference*. Timber Press, Inc., 2020.
Cavanaugh, Sarah Rose. *The Spark of Learning: Energizing the Classroom through the Science of Emotion*. West Virginia University Press, 2016.
Ceccarelli, Leah. *Shaping Science with Rhetoric: The Cases of Dobzhansky, Schrodinger, and Wilson*. University of Chicago Press, 2001.
Chaiklin, Seth. "The Zone of Proximal Development in Vygotsky's Analysis of Learning and Instruction." *Vygotsky's Educational Theory and Practice in Cultural Context*, edited by Alex Kozulin, Boris Gindis, Vladimir Ageyev, and Suzanne Miller, Cambridge University, 2003, pp. 39–64.
Chiang, Ted. *Stories of Your Life and Others*. Tor Books, 2002.
Chmielewski, Michael, and Sarah C. Kucker. "An MTurk Crisis? Shifts in Data Quality and the Impact on Study Results." *Social Psychological and Personality Science*, vol. 11, no. 4, 2020, pp. 464–73.
Churchland, Paul. *The Engine of Reason, the Seat of the Soul: A Philosophical Journey into the Brain*. MIT Press, 1999.
Clareson, Thomas, and Joe Sanders. *The Heritage of Heinlein*. McFarland & Company, 2014.
Clark, Timothy. *Ecocriticism on the Edge*. Bloomsbury, 2015.
Clarke, Arthur C. *The Collected Stories of Arthur C. Clarke*. Tor, 2000.
———. *2001: A Space Odyssey*. Signet, 1968.
———. *2010: Odyssey Two*. Del Rey, 1982.
Clute, John. "Gene Wolfe Obituary." *The Guardian*. 29 April 2019, http://www.theguardian.com/books/2019/apr/29/gene-wolfe-obituary.
Coates, Ta-Nehisi. *Between the World and Me*. Spiegel & Grau, 2015.
———. *Black Panther: Nation under Our Feet*. Vol. 3, Marvel, 2016.
———. *Black Panther: The Intergalactic Empire of Wakanda: Book Two*. Marvel, 2018.
Crowdsourcing and Citizen Science Act. United States Congress, November 30, 2016, www.congress.gov/bill/114th-congress/house-bill/6414.
Dalby, Simon. "Framing the Anthropocene: The Good, the Bad, and the Ugly." *The Anthropocene Review*, vol. 3, no. 1, 2016, pp. 31–51.
Davies, Kevin. *The $1,000 Genome: The Revolution in DNA Sequencing and the New Era of Personalized Medicine*. Free Press, 2010.
Davies, Paul. *Eerie Silence*. Houghton Mifflin Harcourt, 2010.

Davis, Lennard. "Biocultures and Education." *Profession*, 2009, pp. 36–43.
Delany, Samuel R. *The Einstein Intersection*. Ace, 1967.
DelliCarpini, Dominic. "Composition, Literary Studies, and the End(s) of Civic Education." *Composition and/or Literature*, edited by Linda S. Bergmann and Edith M. Baker, National Council of Teachers of English, 2006, pp. 17–35.
Dennett, Daniel. *Darwin's Dangerous Idea*. Simon & Schuster, 1995.
D'Erasmo, Stacey. *The Art of Intimacy: The Space Between*. Graywolf Press, 2013.
Dick, Philip K. "'The Double Bill Symposium': Replies to a Questionnaire to Professional SF Writers and Editors (1969)," *The Shifting Realities of Philip K. Dick*, edited by Lawrence Sutin, Vintage Books, 1995.
———. *The Exegesis of Philip K. Dick*, edited by Jonathan Lethem and Pamela Jackson, Gollancz, 2012.
———. *Dr. Bloodmoney, or How We Got Along after the Bomb*. 1965. Mariner Books, 2012.
———. *The Man in the High Castle*. 1962. Mariner Books, 2011.
———. *Scanner Darkly*. 1977. Mariner Books, 2011.
———. *Ubik*. 1969. Mariner Books, 2012.
Doctorow, Cory. *Walkaway: A Novel*. Tor Books, 2017.
Doudna, Jennifer, and Samuel Sternberg. *A Crack in Creation: Gene Editing and the Unthinkable Power to Control Evolution*. Houghton Mifflin Harcourt, 2017.
Dowdall, Lisa. "Treasured Strangers: Race, Biopolitics, and the Human in Octavia E. Butler's *Xenogenesis* Trilogy." *Science Fiction Studies*, vol. 44, no. 3, November 2017, pp. 506–25.
Downs, Doug, and Liane Robertson. "Threshold Concepts in First-Year Composition." *Naming What We Know: Threshold Concepts of Writing Studies*, edited by Linda Adler-Kassner and Elizabeth Wardle, Utah State University Press, 2015, pp. 105–21.
Duckworth, Angela. *Grit: The Power of Passion and Perseverance*. Scribner, 2016.
Dweck, Carol. *Mindset: The New Psychology of Success*. Ballantine Books, 2006.
Eagleton, Terry. *Why Marx Was Right*. Yale University Press, 2011.
Egan, Greg. *The Arrows of Time*. Night Shade Books, 2014.
———. *The Clockwork Rocket*. Night Shade Books, 2012.
———. "Closer." *Axiomatic*. Night Shade, 2014.
———. *Distress*. 1995. Night Shade, 2015.
———. *The Eternal Flame*. Night Shade Books, 2013.
———. *Permutation City*. 1994. Night Shade, 2014.
———. *Zendegi*. 2010. Night Shade, 2011.
Elbow, Peter. *Everyone Can Write: Essays toward a Hopeful Theory of Writing and Teaching Writing*. Oxford University Press, 2000.
Ellis, George F. R. "Issues in the Philosophy of Cosmology." *Philosophy of Physics*, edited by Jeremy Butterfield and John Earman, Elsevier, 2007, pp. 1183–286.
Epstein, Steven. *Impure Science*. University of California Press, 1996.
Epstein, Ziv, Blakeley H. Payne, Judy Hanwen Shen, Abhimanyu Dubey, Bjarke Felbo, Matthew Groh, Nick Obradovich, Manuel Cebrian, and Iyad Rahwan. "Closing the AI Gap." *ArxIV.Org*, 20 March 2018, https://arxiv.org/abs/1803.07233v1.
Eshun, Kodwo. *More Brilliant than the Sun*. Quartet Books, 1998.
Fahnestock, Jeanne. "Accommodating Science: The Rhetorical Life of Scientific Facts." *Written Communication*, vol. 3, no. 3, 1986, pp. 275–96.
———. *Rhetorical Figures in Science*. Oxford University Press, 1999.
Farmer, Frank M., and Phillip K. Arrington. "Apologies and Accommodations: Imitation and the Writing Process." *Rhetoric Society Quarterly*, vol. 23, no. 1, 1993, pp. 12–34.
Farrier, David. *Anthropocene Poetics: Deep Time, Sacrifice Zones, and Extinction*. University of Minnesota Press, 2019.
Fodor, Jerry. *The Mind Doesn't Work That Way: The Scope and Limits of Computational Psychology*. MIT Press, 2000.
Foss, Sonja, and Cindy Griffin. "Beyond Persuasion: A Proposal for an Invitational Argument." *Communications Monographs*, vol. 62, 1995, pp. 2–18.

Franklin, H. Bruce. *Robert Heinlein: America as Science Fiction*. Oxford University Press, 1980.
Franko, Carol. "What-if-ing the Titans: Nancy Kress's Dialogic *Beggars* Trilogy." *Extrapolation*, vol. 43, no. 2, pp. 131–62.
Frey, Renea. "Rhetorics of Reflection: Revisiting Listening Rhetoric through Mindfulness, Empathy, and Nonviolent Communication." *Journal for the Assembly of Expanded Perspectives on Learning*, vol. 23, 2017–2018, pp. 92–104.
Fry, Hannah. *Hello World: Being Human in an Age of Algorithms*. W. W. Norton & Company, 2018.
Funtowicz, Silvio, and Jerome Ravetz. "Science for the Post-Normal Age." *Futures*, vol. 31, no. 7, 1993, 735–55.
Gene-Rowe, Francis. "Uncursing the Cursed World: Fate & Apostolic Reading in Philip K. Dick." *Foundation: The International Review of Science Fiction*, vol. 48, no. 2, July 2019, pp. 60–73.
Ghosh, Amitov. *The Great Derangement: Climate Change and the Unthinkable*. University of Chicago Press, 2016.
Giddens, Anthony. *The Politics of Climate Change*. Polity, 2009.
Glasser, Matthew, Stephen Smith, Daniel Marcus, Jesper Andersson, Edward Auerbach, Timothy Behrens, Timothy Coalson, Michael Harms, Mark Jenkinson, Steen Moeller, Emma Robinson, Stamatios Sotiropoulos, Junqian Xu, Essa Yacoub, Kamil Ugurbil, and David Van Essen. "The Human Connectome Project's Neuroimaging Approach." *Nature*, vol. 19, no. 9, September 2016, pp. 1175–86.
Goonan, Kathleen Ann. "The Future of Identity: Implications, Challenges, and Complications of Human/Machine Consciousness." *Intelligence Unbound: The Future of Uploaded and Machine Minds*. Wiley & Sons, Inc., 2014.
Gordon, Joan. *Gene Wolfe*. Wildside Press, 1986.
Graff, Gerald. "Why How We Read Trumps What We Read." *Profession*, 2009, pp. 66–74.
Gray Matter. Windows Version, Wizardbox, 2010.
Greenemeier, Larry. "Video Game Expands the Concept of Dark Energy for Mass Effect." *Scientific American*, 25 January 2010, www.scientificamerican.com/article/dark-energy-mass-effect.
Griffiths, Paul, and Karola Stotz. *Genetics and Philosophy: An Introduction*. Cambridge University Press, 2013.
Grimes. "We Appreciate Power." *Miss Anthropocene*, 4AD, 2020.
Gross, Alan G. "Rhetoric of Science without Constraints." *Rhetorica*, vol. 9, no. 4, Autumn 1991, pp. 283–99.
Hamilton, Clive. "Getting the Anthropocene So Wrong." *The Anthropocene Review*, vol. 2, no. 2, 2015, pp. 102–7.
Hannibal, Mary. *Citizen Scientist: Searching for Heroes and Hope in an Age of Extinction*. The Experiment, 2016.
Haraway, Donna. "Anthropocene, Capitalocene, Plantationocene, Chthulucene: Making Kin." *Environmental Humanities*, vol. 6, 2015, pp. 159–65.
Harwit, Martin. *Cosmic Discovery: The Search, Scope, and Heritage of Astronomy*. MIT Press, 1984.
Hayles, N. Katherine. "Greg Egan's *Quarantine* and *Teranesia*: Contributions to the Millennial Reassessment of Consciousness and the Cognitive Nonconscious." *Science Fiction Studies*, vol. 42, no. 1, March 2015, pp. 56–77.
———. "Hyper and Deep Attention: The Grenerational Divid in Cognitive Modes." *Profession*, 2007, 187–99.
Heinlein, Robert. *Methuselah's Children*. 1958. Signet, 1960.
———. "On the Writing of Speculative Fiction." *The Nonfiction of Robert Heinlein: Volume 1*. The Virginia Edition, 2011.
———. *Time Enough for Love*. Berkley Medallion Books, 1973.
Heller, Jason. *Strange Stars: David Bowie, Pop Music, and the Decade Sci-Fi Exploded*. Melville House, 2018.

Herbert, Frank. *Destination: Void*. Berkley Medallion, 1966.
Herndl, Carl. "Cultural Studies and Critical Science." *Understanding Scientific Prose*, edited by Jack Selzer, The University of Wisconsin Press, 1993, pp. 61–81.
Higgins, David. "*Orphan Black* and Neocitizenship." *Science Fiction Film and Television*, vol. 11, no. 3, 2018, pp. 519–20.
———. "Toward a Cosmopolitan Science Fiction." *American Literature*, vol. 83, no. 2, 2011, pp. 331–54.
Hofstadter, Richard. *Social Darwinism in American Thought, 1860–1915*. University of Pennsylvania Press, 1944.
Hogan, Patrick Colm. *Cognitive Science, Literature, and the Arts: A Guide for Humanists*. Routledge, 2003.
hooks, bell. *Teaching to Transgress: Education as the Practice of Freedom*. Routledge, 1994.
Hulme, Tim. *Why We Disagree About Climate Change: Understanding Controversy, Inaction, and Opportunity*. Cambridge University Press, 2009.
Huron, David. *Sweet Anticipation*. A Bradford Book, 2008.
Huxley, Aldous. *Brave New World*. 1932. Harper Perennial Classics, 2005.
"I Am Also a We." *Sense8*. Created by Lana Wachowski and Lilly Wachowski, season 1, episode 2, Netflix, 2015.
Ioannidis, John. "Why Most Published Research Findings Are False." *Chance*, vol. 18, 2005, pp. 40–47.
Irwin, James. *Citizen Science: A Study of People, Expertise, and Sustainable Development*. Routledge, 1995.
Jameson, Fredric. *Archaeologies of the Future*. Verso, 2005.
———. "World-Reduction in Le Guin: The Emergence of Utopian Narrative." *Modern Critical Views: Ursula K. Le Guin*, pp. 57–70.
Jasanoff, Sheila. *Designs on Nature: Science and Democracy in Europe and the United States*. Princeton University Press, 2007.
Jemisin, N. K. *The Fifth Season*. Orbit, 2015.
Joyce, James. *Portrait of the Artist as a Young Man*. B. W. Huebsch, 1916.
Kaeberlein, Matt. "The Biology of Aging: Citizen Scientists and Their Pets as a Bridge between Research on Model Organisms and Human Subjects." *Veterinary Pathology*, vol. 53, no. 2, 2016, pp. 291–98.
Kakoudaki, Despina. *An Anatomy of a Robot*. Rutgers University Press, 2014.
Kaku, Michio. *Visions: How Science Will Revolutionize the 21st Century*. Anchor, 1998.
Keen, Suzanne. *Empathy and the Novel*. Oxford University Press, 2007.
Keller, Daniel. *Chasing Literacy: Reading and Writing in an Age of Acceleration*. Utah State University Press, 2013.
Kellert, Stephen. "The Biological Basis for Human Values of Nature." *The Biophilia Hypothesis*, edited by Stephen Kellert and Edward O. Wilson, Island Press, 2013.
Kelty, Christopher. *Two Bits: The Cultural Significance of Free Software*. Duke University Press, 2008.
Kilgore, De Witt Douglas. *Astrofuturism: Science, Race, and Visions of Utopia in Space*. University of Pennsylvania Press, 2003.
King, Tom. *Vision*. Marvel, 2018.
Kirby, David. *Lab Coats in Hollywood: Science, Scientists, and Cinema*. MIT Press, 2011.
Klein, Naomi. *This Changes Everything*. Simon & Shuster, 2014.
Knight, Damon. *In Search of Wonder: Essays on Modern Science Fiction*. Advent Publishers, 1967.
Kochenov, Dimitry. *Citizenship*. MIT Press, 2019.
Koerner, Brendan. "New Videogame Lets Amateur Researchers Mess with RNA." *Wired Magazine*. 5 July 2012.
Kostov, V. B., P. R. McCullough, T. C. Hinse, Z. I. Tsvetanov, G. Hebrard, R. F. Dıaz, M. Deleuil, and J. A. Valenti. "A Gas Giant Circumbinary Planet Transiting the F Star Primary of the Eclipsing Binary Star KIC 4862625 And the Independent Discovery and Characterization of Two Transiting Planets in the KEPLER-47 System." *The Astrophysical Journal*, vol. 770, no. 52, 2013, pp. 1–19.

Kozubek, Jim. *The Modern Prometheus: Editing the Human Genome with CRISPR-Cas9*. Cambridge University Press, 2016.

Kress, Nancy. *If Tomorrow Comes*. Tor Books, 2018.

———. *Terran Tomorrow*. Tor Books, 2018.

———. *Yesterday's Kin*. Tachyon Publications, 2014.

Kuhn, Thomas. *The Structure of Scientific Revolutions*. The University of Chicago Press, 1962.

Lang, James. *Small Teaching: Everyday Lessons from the Science of Learning*. Jossey-Bass, 2016.

Latour, Bruno. "Where Are the Missing Masses? The Sociology of Mundane Artifacts." *Technology and Society, Building Our Sociotechnical Future*, edited by Deborah J. Johnson and Jameson M. Wetmore, MIT Press, 2008, pp. 151–80.

Le Guin, Ursula K. *Hainish Novels & Stories: Volume One*, edited by Brian Attebery, The Library of America, 2017.

———. "Do-It-Yourself Cosmology." *The Language of the Night: Essays on Fantasy and Science Fiction*, edited by Susan Wood, G. P. Putnam's Sons, 1979, pp. 121–25.

Lemov, Rebecca. *World as Laboratory: Experiments with Mice, Mazes, and Men*. Hill and Wang, 2005.

Lessl, Thomas. *Rhetorical Darwinism: Religion, Evolution, and the Scientific Identity*. Baylor University Press, 2012.

Lilienfeld, Scott, and Irwin Waldman. *Psychological Science under Scrutiny: Recent Challenges and Proposed Solutions*. Wiley & Sons, Incorporated, 2017.

Lindemann, Erika. "Freshman Composition: No Place for Literature." *College English*, vol. 55, no. 3, March 1993, pp. 311–16.

Lintott, Chris. *The Crowd and the Cosmos*. Oxford University Press, 2019.

Liu, Cixin. *Ball Lightning*. Translated by Joel Martinsen, Tor, 2019.

Loarie, Scott R., Benjamin E. Carter, Katharine Hayhoe, Sean McMahon, Richard Moe, Charles A. Knight, and David D. Ackerly. "Climate Change and the Future of California's Endemic Flora." *PLoS ONE*, vol. 3, no. 6, 2008.

Luckhurst, Roger. *Alien*. British Film Institute, 2014.

———. *The Invention of Telepathy*. Oxford University Press, 2002.

MacLeod, Ken. *The Corporation Wars Trilogy*. Orbit, 2018.

Markley, Robert. *Kim Stanley Robinson*. University of Illinois Press, 2019.

Mastroianni, Brian. "Dog Aging Study Helps to Extend Lifespan." *CBS News*, 19 May 2016.

Mayer, John D., and Peter Salovey. "What Is Emotional Intelligence?" *Emotional Development and Emotional Intelligence*. Basic Books, 1997, pp. 3–31.

Mayor, Michel, and Didier Queloz. "A Jupiter-Mass Companion to a Solar-Type Star." *Nature*, vol. 378, 1995, pp. 355–59.

McCormick, Kathleen. "Closer than Close Reading: Historical Analysis, Cultural Analysis, and Symptomatic Reading in the Undergraduate Classroom." *Intertexts: Reading Pedagogy in College Writing Classrooms*, edited by Marguerite Helmers, Routledge, 2002, pp. 26–47.

———. *The Culture of Reading/The Teaching of English*. Manchester University Press, 1994.

McGuirk, Carol. "Optimism and the Limits of Subversion in *The Dispossessed* and *The Left Hand of Darkness*." *Modern Critical Views: Ursula K. Le Guin*, edited by Harold Bloom, Chelsea House Publisher, 1986.

Mendlesohn, Farah. *The Pleasant Profession of Robert Heinlein*. Unbound, 2019.

Meyer, Jan H. F., and Ray Land. "Threshold Concepts and Troublesome Knowledge: Issues of Liminality." *Overcoming Barriers to Student Understanding: Threshold Concepts and Troublesome Knowledge*, edited by Jan H. F. Meyer and Ray Land, Routledge, 2006, pp. 19–32.

Miéville, China. *Embassytown: A Novel*. Del Rey, 2011.

———. "Afterword: Cognition as Ideology: A Dialectic of SF Theory." *Red Planets: Marxism and Science Fiction*, edited by Mark Bould and China Miéville, Pluto Press, 2009.

Milburn, Colin. *Mondo Nano: Fun and Games in the World of Digital Matter*. Duke University Press, 2015.

Miller, Sam J. *Blackfish City: A Novel*. HarperCollins Publishers, 2018.

———. "Q&A with *Blackfish City* Novelist Sam Miller," by Amy Brady. *Yale Climate Connections*, 6 June 2019. Accessed 23 October 2020.

Mitchell, Sara. "Exploring the Galaxy with Mass Effect 3." *Blueshift*, 31 July 2012. Accessed 23 October 2020.

Mooney, Chris, and Sheril Kirshenbaum. *Unscientific America: How Scientific Illiteracy Threatens Our Future*. Basic Books, 2009.

Morgan, Patricia Fey. "A Brief History of the Current Re-emergence of Contemplative Education." *Journal of Transformative Education*, vol. 13, no. 3, 2015, pp. 197–215.

Morrow, Nancy. "The Role of Reading in the Composition Classroom." *Journal of Advanced Composition*, vol. 17, no. 3, 1997, pp. 453–72.

Morton, Timothy. *Dark Ecology: For the Logic of Coexistence*. Columbia University Press, 2016.

———. *Hyperobjects: Philosophy and Ecology at the End of the World*. University of Minnesota Press, 2013.

Mukherjee, Siddhartha. *The Gene: An Intimate History*. Scribner, 2016.

Nama, Adilifu. *Super Black: American Pop Culture and Black Superheroes*. University of Texas Press, 2011.

Nascimento, Susana, Jose Miguel Rubio Iglesias, Roger Owen, Sven Schade, and Lea Shanley. "Citizen Science for Policy Formation and Implementation." *Citizen Science: Innovations in Open Science, Society, and Policy*, pp. 219–40.

National Academies of Science. "Learning through Citizen Science: Enhancing Opportunities by Design." National Academies Press, 2019.

Nelkin, Dorothy, and M. Susan Lindee. "Cloning in the Popular Imagination." *Cloning Sourcebook*, edited by Arlene Judith Klotzko, Oxford University Press, 2003, pp. 83–93.

Newitz, Annalee. *Autonomous*. Tor, 2017.

———. "When Robot and Crow Saved East St. Louis." *Slate*. December 2018, slate.com/technology/2018/12/annalee-newitz-short-story-when-robot-and-crow-saved-east-st-louis.html.

Nielsen, Michael. *Reinventing Discovery: The New Era of Networked Science*. Princeton University Press, 2011.

Noble, Safiya Umoja. *Algorithms of Oppression*. New York University Press, 2018.

Nogués-Bravo, David, Daniel Simberloff, Carsten Rahbek, and Nathan James Sanders. "Rewilding is the New Pandora's Box in Conversation." *Current Biology*, vol. 26, February 2016, 87–91.

Nussbaum, Martha. *Cultivating Humanity: A Classical Defense of Reform in Liberal Education*. Harvard University Press, 1998.

Oatley, Keith. *The Best-Laid Schemes: The Psychology of Emotions*. Cambridge University Press, 1992.

O.K. Computer. Radiohead. Capitol, 1997.

O'Neil, Cathy. *Weapons of Math Destruction*. Broadway, 2016.

Okorafor, Nnedi. *Binti*. Tor, 2015.

———. *Binti: Home*. Tor, 2017.

———. *Binti: The Night Masquerade*. Tor, 2017.

Overington, Michael A. "The Scientific Community as Audience: Toward a Rhetorical Analysis of Science." *Philosophy and Rhetoric*, vol. 10, no. 3, 1977, pp. 143–64.

Page, Michael. *Frederik Pohl*. University of Illinois Press, 2015.

"Paradigm Shift." *The Expanse*, written by Ty Frank, Daniel Abraham, and Naren Shankar, directed by David Grossman, *SyFy*, 2017.

Parsemain, Ava Laure. *The Pedagogy of Queer TV*. Palgrave MacMillan, 2019.

Passmore, John. *Science and Its Critics*. Duckworth, 1978.

Peary, Alexandria. "Mindfulness, Buddhism, and Rogerian Argument." *Journal for the Assembly for Expanded Perspective on Learning*, vol. 11, 2006, pp. 64–75.

Perrow, Charles. *Normal Accidents: Living with High-Risk Technologies*. Basic Books, 1984.

Phillips, Kaitlin Ugolik. *The Future of Feeling: Building Empathy in a Tech-Obsessed World*. Little A, 2020.

Phillips, Tina B., Heidi Ballard, Bruce Lewenstein, and Rick Bonney. "Engagement in Science through Citizen Science: Moving beyond Data Collection." *Science Education*, vol. 103, no. 1, 2019, pp. 665–90.

Pickover, Clifford. *The Science of Aliens*. Basic Books, 1998.

Pink Floyd. *The Dark Side of the Moon*. Harvest, 1973.

Pohl, Frederik. *The Merchants' War*. St. Martin's Press, 1984.

Pohl, Frederik, and C. M. Kornbluth. *The Space Merchants*. Del Rey, 1952.

Polyani, Michael. "The Republic of Science: Its Political and Economic Theory." *Minerva*, vol. 1, no. 1, Autumn 1962, 54–73.

Porter, Roy. *Madness: A Brief History*. Oxford University Press, 2003.

Powers, Richard. *Galatea 2.2: A Novel*. Harper Perennial, 1995.

Prainsack, Barbara. "Understanding Participation: The 'Citizen Science' of Genetics." *Genetics as Social Practice*, edited by Barbara Prainsack, Gabriele Werner-Felmayer, and Silke Schicktanz, Routledge, 2014, pp. 146–64.

Pyle, Rod. *Space 2.0: How Private Spaceflight, a Resurgent NSA, and International Partners Are Creating a New Space Age*. Ben Bella Books, 2019.

Qualley, Donna. *Turns of Thought*. Boynton/Cook, 1997.

Rabkin, Eric S. "Determinism, Free Will, and Point of View in *The Left Hand of Darkness*." *Modern Critical Views: Ursula K. Le Guin*, pp. 155–70.

———. "Immortality: The *Self*-Defeating Fantasy." *The Immortal Engines: Life-Extension and Immortality in Science Fiction and Fantasy*, edited by George Edgar Slusser, Gary Westfahl, and Eric Rabkin, University of Georgia Press, 1996, pp. ix–xvii.

Raddick, M. Jordan, Georgia Bracey, Pamela L. Gay, Chris J. Lintott, Phil Murray, Kevin Schawinski, Alexander S. Szalay, and Jan Vandenberg. "Exploring the Motivations of Citizen Science Volunteers." *Astronomy Education Review*, 2010, vol. 9, no. 1.

Reardon, Jenny. *Race to the Finish: Identity and Governance in an Age of Genomics*. Princeton University Press, 2004.

Richter, Anette, Daniel Dörler, Susanne Hecker, Florian Heigl, Lisa Pettibone, Fermin Serrano Sanz, Katrin Vohland, and Aletta Bonn. "Capacity Building in Citizen Science." *Citizen Science: Innovation in Open Science, Society, and Policy*. UCL Press, 2018, pp. 269–83.

Rieder, John. *Science Fiction and the Mass Cultural Genre System*. Wesleyan University Press, 2017.

Roberts, Dorothy. *Fatal Invention: How Science, Politics, and Big Business Re-create Race in the Twenty-First Century*. The New Press, 2011.

Robinson, Kim Stanley. *2312*. Orbit, 2012.

———. *Aurora*. Orbit, 2015.

———. *The Lucky Strike*. PM Press, 2009.

———. *The Novels of Philip K. Dick*. UMI Research Press, 1984.

Robson, Justina. *Mappa Mundi*. Pyr, 2006.

Rood, Craig. "Rhetorics of Civility: Theory, Pedagogy, and Practice in Speaking and Writing Textbooks." *Rhetoric Review*, vol. 32, no. 3, 2013, 331–48.

Rose, Phil. *Roger Waters and Pink Floyd: The Concept Albums*. Farleigh Dickinson University Press, 2015.

Ross, Alec. *The Industries of the Future*. Simon and Shuster, 2016.

Rothman, Joshua. "The Weird Thoreau." *The New Yorker*, 14 January 2015.

Rubin, Daniel Ian. "Mindcrime and Doublethink: Using Music to Teach Dystopian Literature." *The English Journal*, vol. 101, no. 2, November 2011, pp. 74–79.

Salzberg, Sharon. *Real Love: The Art of Mindful Connection*. Flatiron Books, 2017.

Sanders, Lisa. "In the Heat of the Night." *New York Times Magazine*, 12 September 2010, pp. 26–28.

Schmeink, Lars. *Biopunk Dystopias*. Liverpool University Press, 2016.

Scholes, Robert. *Protocols of Reading*. Yale University, 1989.

Seager, Sara. "Exoplanet Habitability." *Science Magazine*, vol. 340, 2013, pp. 577–81.

Sègla, Aimé Dafon. "The Cosmological Vision of the Yoruba-Id` a`acha of Benin Republic (West Africa): A Light on Yoruba History and Culture." *African Cultural Astronomy*, edited by Jarita Holbrook, R. Thebe Medupe, and Johnson O. Urama, Springer, 2008, pp. 289–208.

Selzer, Jack. *Understanding Scientific Prose*. The University of Wisconsin Press, 1993.

Seung, Sebastian. *Connectome: How the Brain's Wiring Makes Us Who We Are*. Houghton Mifflin Harcourt, 2012.

Shanahan, Mike. "Talking About a Revolution: Climate Change and the Media." *International Institute for Environment and Development*, December 2007.

Shane, Janelle. "No Robot like Robot." *Slate*, December 2018, slate.com/technology/2018/12/machine-learning-ai-annalee-newitz-when-robot-and-crow-saved-east-st-louis.html.

———. *You Look like a Thing and I Love You*. Voracious, 2019.

Shapin, Steven. *The Scientific Revolution*. University of Chicago Press, 1996.

Shelley, Mary Wollstonecraft. *Frankenstein; Or, the Modern Prometheus*. 1818. Ed. Susan J. Wolfson. Longman, 2007.

Smith, Philip E. "The Evolution of Politics and the Politics of Evolution: Social Darwinsim in Heinlein's Fiction." *Robert A. Heinlein*, edited by Joseph Olander and Martin Harry Greenberg, Taplinger Publishing Co., 1978, pp. 138–42.

Snaza, Nathan. "The Earth is Not 'Ours' to Save." *Interrogating the Anthropocene: Ecology, Aesthetics, Pedagogy, and the Future in Question*, edited by Jan Jagodzinski, Palgrave Macmillan, 2019.

Sorenson, Karen Schroeder. Cosmos *and the Rhetoric of Popular Science*. Lexington Books, 2017.

Soulé, Michael, and John Terborgh. "The Policy and Science of Regional Conversation."*Continental Conversation: Scientific Foundations of Regional Reserve Networks*. Island Press, 1999.

Sousa, David A. *How the Brain Learns*. Corwin, 2011.

Squarzoni, Philippe. *Climate Changed: A Personal Journey through the Science*. Harry N. Abrams, 2014.

Stableford, Brian. *The Fountains of Youth*. Tor Books, 2005.

Star Trek: First Contact. Dir. Jonathan Frakes. Perf. Patrick Stewart, Jonathan Frakes, Brent Spiner. Paramount, 1996.

Starkey, Natalie. *Catching Stardust: Comets, Asteroids, and the Birth of the Solar System*. Bloomsbury Publishing, 2018.

Stephenson, Neal. *The Diamond Age*. Bantam Spectra, 1995.

Suvin, Darko. "Artifice as Refuge and World View: Philip K. Dick's Foci." *Philip K. Dick*, edited by Martin Greenberg and Joseph Olander, New York, 1983, pp. 73–95.

Swan, Judith. "Reflections across the Divide: Written Discourse as a Structural Mirror in Teaching Science to Nonscience Students." *Writing on the Edge*, vol. 6, no. 2, Spring 1995, 55–73.

Tang, Yi-Yuan, Britta K. Hölzel, and Michael I. Posner. "The Neuroscience of Mindfulness Meditation." *Nature*, vol. 16, 2015, pp. 213–25.

Tate, Gary. "A Place for Literature in Freshman Composition." *College English*, vol. 55, no. 3, March 1993, 317–21.

Taylor, Charlotte. "Threshold Concepts in Biology: Do They Fit the Definition?" *Overcoming Barriers to Student Understanding*, pp. 87–89.

Tinati, Ramine, Markus Luczak-Roesch, Elena Simperl, and Wendy Hall. "An Investigation of Gamers' Motivations in *EyeWire*, a Gamified Citizen Science Project." *Computers in Human Behavior*, vol. 73, 2017, pp. 527–48.

Tobin, Vera. *Elements of Surprise: Our Mental Limits and Satisfactions of Plots*. Harvard University Press, 2018.
Tompkins, David. "Weird Ecology: *The Southern Reach* Trilogy." *Los Angeles Review of Books*, 30 September 2014.
Tompkins, Jane. *Reader Response Criticism: From Formalism to Poststructuralism*. Johns Hopkins University Press, 1980.
Trexler, Adam. *Anthropocene Fiction: The Novel in a Time of Climate Change*. University of Virginia Press, 2015.
Tribbe, Matthew. *No Requiem for the Space Age: The Apollo Moon Landings in American Culture*. Oxford University Press, 2014.
Ulstein, Gry. "Brave New Weird: Anthropocene Monsters in Jeff VanderMeer's *The Southern Reach*." *Concentric: Literary and Cultural Studies*, vol. 43, issue 1, 2017, pp. 71–96.
Urama, Johnson, and Jarita Holbrook. "The African Cultural Astronomy Project." *Proceedings International Astronomical Union Symposium*, no. 260, 2009, pp. 48–53.
Vaidhyanathan, Siva. *Antisocial Media: How Facebook Disconnects Us and Undermines Democracy*. Oxford University Press, 2018.
VanderMeer, Jeff. *Annihilation*. Farrar, Straus, and Giroux, 2014.
Vermeule, Blakey. *Why Do We Care about Literary Characters?* Johns Hopkins University Press, 2009.
Vertesi, Janet. *Seeing like a Rover: How Robots, Teams, and Images Craft Knowledge of Mars*. University of Chicago Press, 2015.
Vincent, Rickey. *Funk: The Music, the People, and the Rhythm of the One*. St. Martin's Publishing Group, 1996.
Vint, Sherryl. *Bodies of Tomorrow: Technology, Subjectivity, Science Fiction*. University of Toronto Press, 2007.
Wagner, John. "Full Earth Crimes." *Judge Dredd: The Complete Case Files 01*. Rebellion, 2013.
Wagner, Sven. *The Scientist as God: A Typological Study of a Literary Motif, 1818 to the Present*. University of Heidelberg, 2012.
Wark, McKenzie. *Gamer Theory*. Harvard University Press, 2009.
Webster-Wright, Ann. "The Eye of the Storm: A Mindful Inquiry into Reflective Practices in Higher Education." *Reflective Practice*, vol. 14, no. 4, pp. 556–67.
Weisberg, Deena Skolnick, Frank C. Keil, Joshua Goodstein, Elizabeth Rawson, and Jeremy R. Gray. "The Seductive Allure of Neuroscience Explanations." *Journal of Cognitive Neuroscience*, vol. 20, no. 3, 2008, pp. 470–77.
Wells, Herbert George. *The Island of Doctor Moreau*. 1896. Penguin Classics, 2005.
Wilson, Edward O. *Biophilia*. Harvard University Press, 1984.
Wolfe, Gene. "From a House on the Borderland." *Castle of Days*. Macmillan, 1995.
———. *Shadow and Claw*. Orb, 1994.
———. *Sword and Citadel*. Orb, 1994.
Wolf, Maryanne. *Reader, Come Home: The Reading Brain in a Digital World*. Harper, 2018.
Womack, Ytasha. *Afrofuturism: The World of Black Sci-Fi and Fantasy Culture*. Lawrence Hill Books, 2013.
Wright, Peter. *Attending Daedalus: Gene Wolfe, Artifice, and the Reader*. Liverpool University Press, 2003.
Wynn, James. *Citizen Science in the Digital Age: Rhetoric, Science and Public Engagement*. University of Alabama Press, 2017.
Yeager, David, Marlone Henderson, David Paunesku, Gregory M. Walton, Sidney D'Mello, Brian J. Spitzer, and Angela Duckworth. "Boring but Important: A Self-Transcendent Purpose for Learning Fosters Academic Self-Regulation." *Journal of Personality and Social Psychology*, vol. 107, no. 4, pp. 559–80.
Zajonc, Arthur. "Contemplative Pedagogy in Higher Education toward a More Reflective Academy." *Contemplative Learning and Inquiry across Disciplines*, edited by Olen Gunnlaugson, Edward W. Sarath, Charles Scott, and Heesoon Bai, State University of New York Press, 2014.

Zerbe, Michael. *Composition and the Rhetoric of Science: Engaging the Dominant Discourse.* Southern Illinois University, 2007.

Zhao, Cecilia Guanfang. "Voice in Timed L2 Argumentative Essay Writing." *Assessing Writing,* vol. 31, no. 1, 2017, pp. 73–83.

Zunshine, Lisa. *Why We Read Fiction: Theory of Mind and the Novel.* Ohio State University Press, 2006.

Index

About the Author

Jerome Winter, PhD, is a full-time lecturer at the University of California, Riverside. His first book, *Science Fiction, New Space Opera, and Neoliberal Globalism* (2016), was published by the University of Wales Press as part of their New Dimensions in Science Fiction series. His various scholarship has appeared in *The Oxford Handbook of Science Fiction, Extrapolation, Journal of Fantastic and the Arts,* the *Los Angeles Review of Books, Foundation, SFRA,* and *Science Fiction Studies,* among other venues.

www.ingramcontent.com/pod-product-compliance
Ingram Content Group UK Ltd.
Pitfield, Milton Keynes, MK11 3LW, UK
UKHW041842031026
15587UKWH00002B/961

9 781793 621474